Historical Tunnels in the Swiss Alps

Gotthard Simplon Lötschberg

K. Kovári
Professor of Tunneling
Swiss Federal Institute of Technology Zurich

R. Fechtig
Professor emeritus of Construction Engineering and Managements
Swiss Federal Institute of Technology Zurich

Society for the Art of Civil Engineering Volume 2

Thanks are due for their valuable contribution to:

Ch. Amstad
(Swiss Federal Institute of Technology Zurich)

B. Odermatt
(Zurich)

W. Kradolfer
(Rothpletz, Lienhard + Cie. AG, Olten)

M. Sintzel
(Swiss Federal Institute of Technology Zurich)

Original version:
"Historische Alpendurchstiche in der Schweiz"
Gesellschaft für Ingenieurbaukunst, 1996
Translated by Dr. E.G. Prater
© 2000 Society for the Art of Civil Engineering
Printed by: Stäubli AG, Zurich

ISBN 3 7266 0041 8

Front Cover:
Excavation crew in the Lötschberg Tunnel (around 1910)

Contents

▽ The Lötschberg Tunnel at the south ramp: construction of the portal in soft ground.

Foreword

Today we are witnessing the dawn of a new era of railway construction in Switzerland. The required capacity of the planned network and the necessity nowadays to consider ecological factors call for the construction of deep long tunnels through the Alps. The 19 km long Vereina Tunnel has just started operation and the two Base Tunnels Gotthard (57 km) and Lötschberg (35 km) are already under construction. At the centre of these developments, tunnelling as one of the oldest disciplines in civil engineering is faced with great challenges and undreamed of technical possibilities. Besides constructional aspects, political and planning features are also very important in the realisation of these structures.

With this background it is worthwhile casting a look back at the beginnings of Alpine tunnel construction and some of the benchmarks of its development up to the 50s and 60s of the last century. The first major Alpine rail crossing was the 12.2 km long double-track Mont Cenis Tunnel between Italy and France, which was built in 1857–1871. An innovative feature of this project was the mechanisation of blast hole drilling. Based on the compressed air power transmission principle as proposed by the Swiss physicist J.D. Callodon, a drilling carriage was constructed and commissioned by the Italian engineer G. Sommeiller, a pioneer of Alpine tunnelling. The next major tunnel project, the 15 km long double-track Gotthard Tunnel in Switzerland (1872–1881), saw the substitution of gunpowder by dynamite, invented in 1875 by Alfred Nobel. Other Alpine tunnelling projects became famous for overcoming immense geological difficulties, amongst them the Arlberg Tunnel (1880–1884, 10.3 km long in Austria).

The early Alpine railway tunnelling culminated in the 19.8 km long Simplon Tunnel between Italy and Switzerland. The complete structure consists of two parallel single-track tunnels (Simplon I and II) with cross cuts every 200 m. The topographical conditions of the Simplon site, where the overburden reaches 2200 m, meant that tunnelling operations could proceed only from the two portals. With its great length, high rock temperatures (up to 55.4 °C), geological site investigation limitations, and both squeezing and bursting rock conditions, this tunnel is one of the landmarks in the history of tunnelling. Simplon I was built in 1898–1906, Simplon II started operation in 1921. The 14.6 km long Lötschberg Tunnel as a continuation along the south–north railway line was constructed between 1908 and 1913. For almost 70 years, the Simplon was the world's longest railway tunnel until, in 1984, it was surpassed by the 54 km long Seikan Tunnel in Japan. The design and construction reports are associated with the German engineers A. Brandt, K. Brandau and K. Pressel, and Swiss engineers Ch. Andreae, E. Locher, F. Rothpletz, E. Wiesmann, et al. For the first time, squeezing rock was correctly interpreted in relation to the formation of a plastic zone around the underground opening. One of the fundamental features of squeezing rock – i.e. the rock pressure decreases with increased tunnel convergence – was clearly formulated by Wiesmann as follows: "For every fraction of a millimetre of rock deformation, the rock pressure decreases."

The period after World War I was dominated by the construction of hydroelectric power plants, reaching its peak in the 1970s. The significant volume of underground work associated with hydro schemes in the Alps can be measured by the length of water tunnels – more than 10,000 km – and the many rock caverns for underground power stations. This period saw another revolutionary change in tunnel construction technology. Based on developments in the US, timber props were gradually supplemented by steel support systems. During this same

▽ Drilling machine, in Göschenen (Gotthard Railway Tunnel)

period, substantial progress was made in the technology of drilling and blasting, most notably fostered by Swedish scientists, particularily by U. Langefors.

A big technological step in the history of tunnelling was the employment of shotcrete and rockbolting as new ground support elements for tunnels. The first so-called gunite machines were invented in the US and further developed in Germany where, as early as the 1920s and 1930s, they found regular application in both mining and tunnelling. Reinforced gunite lining was applied in a number of tunnels in the UK (e.g. the Mersey Tunnel), which led to the publication of a handbook on Cement Gun Work in London in 1934. Thanks to the refinements and patented inventions of Swiss engineers, shotcrete machines reached the industrial production stage in the 1950s and 1960s. The scene was set for the widespread application of shotcrete as temporary support. The Austrian engineer L. Rabcewicz, wrote in 1964: "The first successful application of surface stabilisation by shotcrete for tunnels in unstable ground as an integral part of the driving process, instead of using timber or steel as temporary support, was carried out at the Lodano-Losogno Tunnel for the Maggia hydroelectric scheme, Switzerland 1951–1955."

The introduction of rockbolting technology for tunnels was also the result of efforts on a broad international scale. Lang (US) concluded in his 1961 state-of-the-art report on rockbolting: "A special tribute must be paid to the US mining industry for its work in initiating and developing the use of rockbolts." Systematic research on the effectiveness of rockbolting was carried out in 1948 by the US Bureau of Mines, in Sweden and in connection with the giant Snowy Mountains Scheme in Australia. In the Alps, systematic rockbolting for tunnelling was initiated during construction of the 11.7 km long pressure tunnel of the Isère Arc hydro scheme in France

(1951–1953). For the first systematic application of shotcrete and rock bolting in a large cross-section traffic tunnel (Schwaikheimer Tunnel, 1963–65) credit is due to the German engineer J. Spang. Widely recognised personalities, like T. A. Lang (USA), A. Sonderegger (Switzerland), F. Mohr (Germany), M. J. Talobre (France), among others, substantially contributed during the 50s and 60s of the last century to the art of tunnelling, leading to the full development of the so-called 'shotcrete tunnelling method', or 'sprayed concrete lining (SCL)' method which in German is referred to as 'Spritzbetonbauweise". Thus tunnelling with shotcrete support and rock bolts and other means of support evolved on a broad international scale.

This book is concerned with the history of four major deep long tunnels through the Swiss Alps having a total length of 87 km: The Gotthard Railway Tunnel, the Simplon Tunnel, the Lötschberg Tunnel and the Gotthard Road Tunnel. The construction of these tunnels falls in the time period between 1872 and 1980.

The original version of this book was published in 1996 in German. It has emerged from the exhibition 'Historical Swiss Alpine Tunnels' which first took place in 1996 at the Museum for the Art of Civil Engineering in Ennenda, Canton Glarus, Switzerland. Further exhibitions were then subsequently organised in several cities in Austria, Germany and Switzerland. The book is now also available in French and Italian.

Zurich February 2000

K. Kovári
R. Fechtig

The Gotthard Railway Tunnel

Total length	14'984 m
Construction period	9 years
Opening date	1st January 1882
Total costs	66.6 mio. SFr.

Artist's view of Göschenen,
1880, wood engraving

History

Overview

1838. First ideas on the construction of a north-south alpine railway through Switzerland. Preference for the passes in the Grisons, especially the Lukmanier route. Expert report of the English engineers R. Stephenson and H. Swinburne in 1850 commissioned by the Swiss Government in favour of the Lukmanier route with an adhesion and cable railway system.

1851. The Winterthur engineer Gottlieb Koller further develops the idea of a Gotthard railway line.

1853. Johann Jakob Speiser, business man and financial expert in Basle and joint founder of the Swiss Central Railway: "By constructing a Gotthard railway line the Central Railway Co. will become international and it can only achieve this by this means." The engineers G. Koller, Winterthur, K. Müller, Altdorf, and P. Lucchini, Locarno, were commissioned to carry out a survey for a future Gotthard route. Application to the Swiss Federal Council by the Cantons Lucerne, Basle City and Basle Land, Nidwalden, Obwalden, Schwyz, Solothurn and Uri for financial and political support.

1863. The Canton Ticino is undecided, but nevertheless commissions an English company to construct the access railway lines Chiasso–Bellinzona–Biasca, both of which would be needed for a future Gotthard or Lukmanier railway line. The project was later adopted and amended by the Gotthard Railway Company in 1872.

1864. Decisive breakthrough in favour of the Gotthard line. The North Eastern Railway Co. under Alfred Escher and thus the Canton Zurich give no support for the Lukmanier route and favour the Gotthard project instead. In 1866 Canton Berne also gave approval to the Gotthard variant and thus put the wellbeing of Switzerland as a whole before its Bernese interests (Member of the Federal Council Stämpfli).

1866. The Cantons Zug, Fribourg, Schaffhausen, Aargau, Thurgau, Neuchâtel and Ticino join the other cantons in favour of the Gotthard project and found with the bigger private railway companies SCB and NOB the "Association of Swiss Cantons and Railway Companies for the Promotion of a Gotthard Railway Line".

1869. Conference in Berne: Italy, Baden-Württemberg and the North German Federation are in favour of the Gotthard project and give assurances of financial support.

1871. 28th October. Signing of the Gotthard Agreement between Italy, the newly formed German Reich and Switzerland. Bismarck: "The best political interests are served by having a connection between Germany and Italy, which is only dependent upon the intermediate neutral country Switzerland, and is not in possession of one of the major European powers."

1871. 6h December: Founding of the Gotthard Railway Company. Chairman of the Board of Directors is Alfred Escher, Zurich. Available capital: 187 million Swiss Francs.

1872. Call for submission of tenders and contract award for the Gotthard Tunnel to the Geneva contractor Louis Favre. Agreed contract sum: 56 million Swiss Francs; agreed construction period: 8 years. Commencement of construction: Autumn 1872.

1875. 28th July: Revolt of the workers in Göschenen because of poor wages and health safety conditions. The use of law and order troops leads to the deaths of four Italian workers. In protest 80 workers walk off the construction site.

1876. Massive excess construction costs for the overall project, international conference to put the project back on its feet financially. Financial help from the Swiss Federal Government, Germany, Italy and the private rail companies NOB and SCB offer loans. Project cuts, provisional dispensing with double track except for the tunnel. Study of variant schemes without the loop tunnels: cable railway, cog railway, etc. Alfred Escher remains firm and defends the original concept.

Emil Welti (1825–1899)
President of the Federal Council

Giovanni B. Pioda (1808–1882)
Member of the Federal Council

Alfred Escher (1819–1882) was a member of the Council of the Canton of Zurich, a member of the National Council, President of the North Eastern Railway Company and Chairman of the Board of Directors of the Gotthard Railway Company founded in 1871.
As a liberal politician he was an ardent champion of private railways in Switzerland. The realisation of the project of the century, the Gotthard line, is mainly thanks to his enormous efforts and his ability to push through a project. He also played a decisive part in the choice of the Gotthard route instead of the proposed Lukmanier variant, whose planning right up to the final decision was well advanced.

The members of the Federal Council Giovanni B. Pioda (1808–1882) and Emil Welti (1825–1899) were the two most important promoters of the Gotthard railway line in the government. At the time of commencement of construction Welti was President of the Federal Council.

1878. After the disputes and due to ill health Alfred Escher withdraws, embittered, into private life.

1879. 19th July: Louis Favre dies of a stroke in the middle of his work. Legal process against his successor, the engineer Bossi, because of cost exceedance. Excessive demands of the Gotthard Railway Company. Despite legally imposed reductions the other demands lead to the subsequent ruin of Favre.

1882. 1st January: Provisional start of railway operation in the tunnel with one track. Completion of the access ramps in spring. Opening celebrations in Lucerne and Milan from the 22nd to the 25th May. From the 1st of June 1882 scheduled timetable of operation on the Gotthard line. On the 6th December Alfred Escher dies in Zurich.

1889. Founding of the Swiss Federal Railways through purchase of the private railways (people's referendum).

1904. Completion of the double track for the most important parts of the Gotthard line.

1909. Gotthard Agreement with Germany and Italy, the Gotthard line taken over by the Swiss Federal Railways.

Geology – Surveying

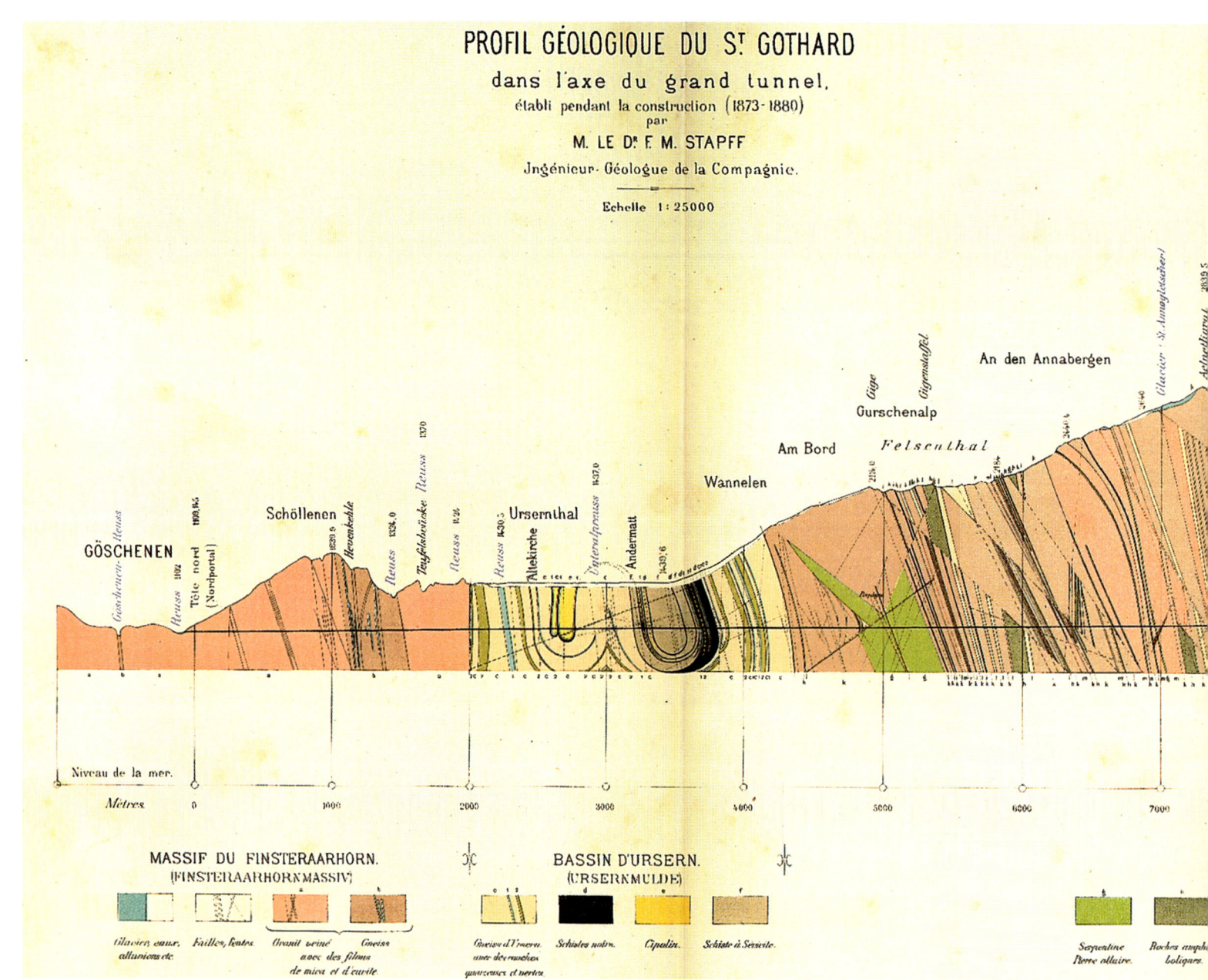

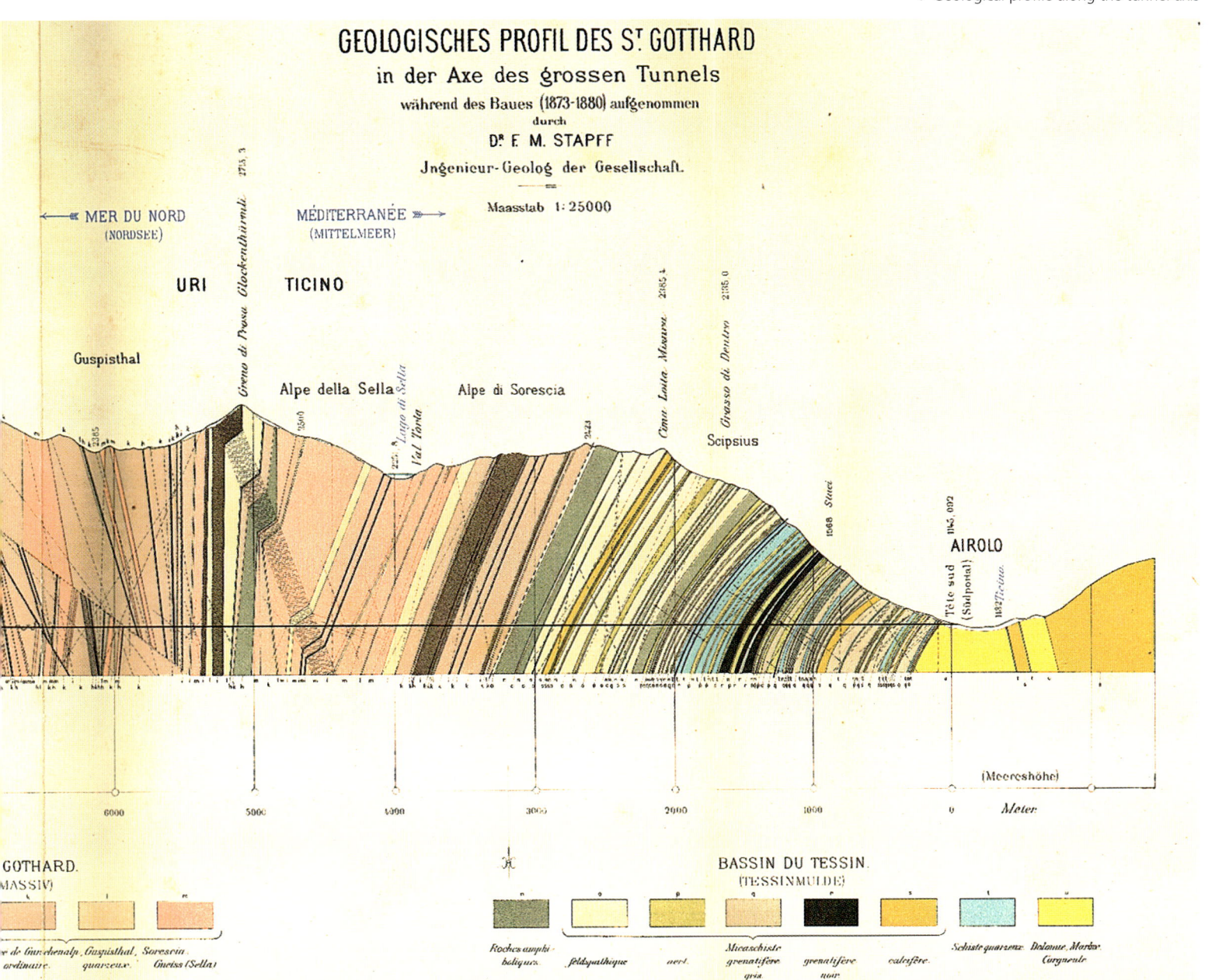

Contracts

In 1872 the Geneva contractor Louis Favre was awarded the contract for the construction of the main tunnel from Airolo to Göschenen. He put in the most favourable offer. Favre had gained wide experience in railway construction and tunnelling – above all in the West of Switzerland and France – and was known to be very reliable. His great optimism may have led him to sign a contract, whose severity would ruin him both physically and financially just seven years later. Favre did not survive to see the completion of his great work. On the 19th July 1879 whilst inspecting the tunnel he died of a stroke. His passing away meant a heavy loss for all those involved in the project, especially for his workers, who had much to thank him for on account of his humane attitude and profound social sensitivity.

▷ Louis Favre
(1826–1879) in his site office

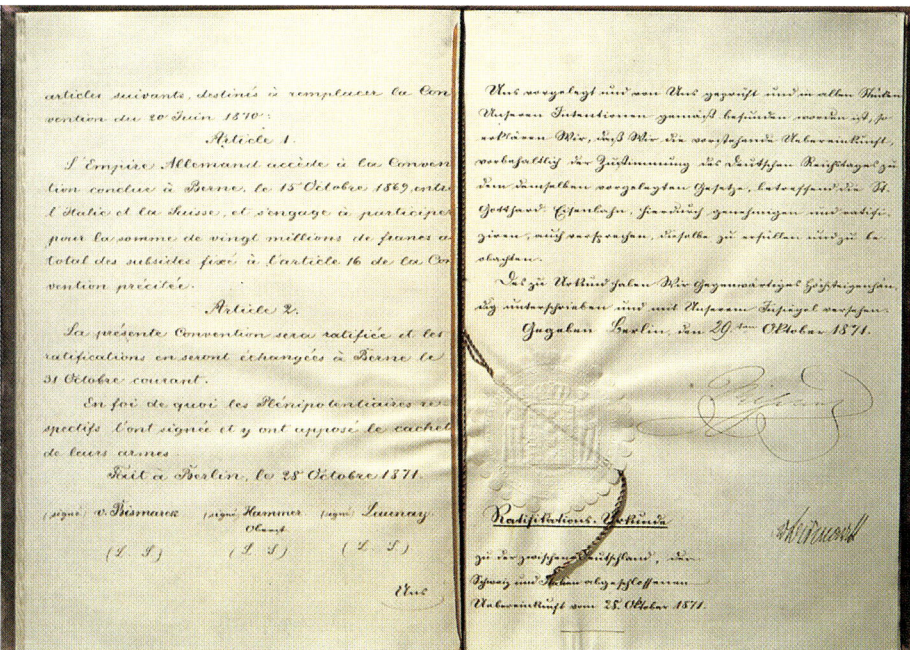

The most significant points of the contract with Louis Favre

Agreed sum of money:
SFr. 55'854'600.–

Construction time: 8 years (exactly to the day)

Depositing SFr. 8 millions in forfait money.

This payment was mainly subscribed by the citizens of Geneva.

Day's premium in his favour for every construction day saved in the period under 8 years: SFr. 5'000.–.

For every day of delay beyond the agreed construction period of 8 years he pledged himself to pay SFr. 5'000.– per day for 6 months, and SFr. 10'000.– after that.

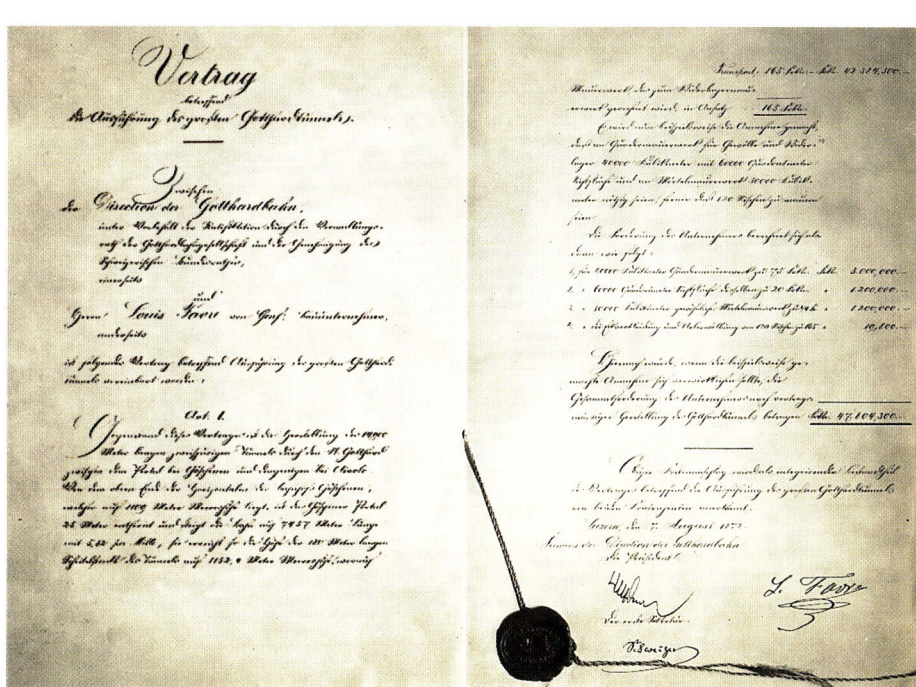

◁ Contract with Louis Favre, first and last pages

Construction Installations

◁ Installation works in Göschenen, view to the south

◁ Inside the compressor building where compressed air was produced for the drilling machines and the tunnel construction railway engines

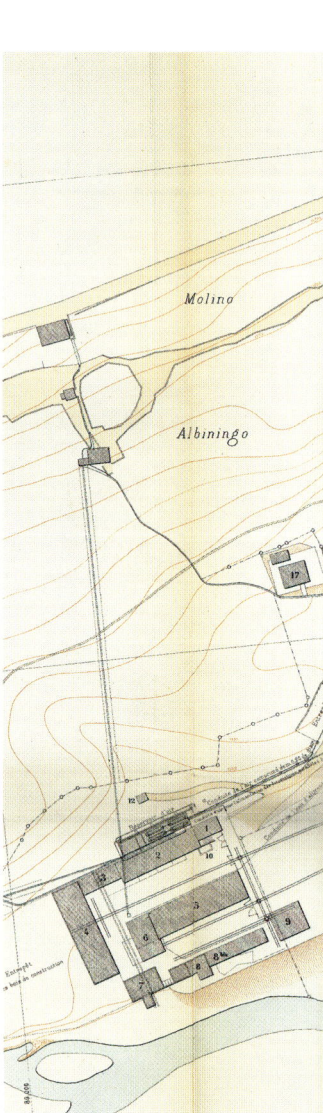

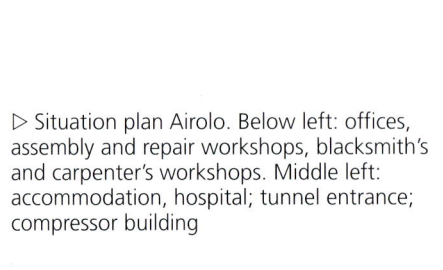

▷ Situation plan Airolo. Below left: offices, assembly and repair workshops, blacksmith's and carpenter's workshops. Middle left: accommodation, hospital; tunnel entrance; compressor building

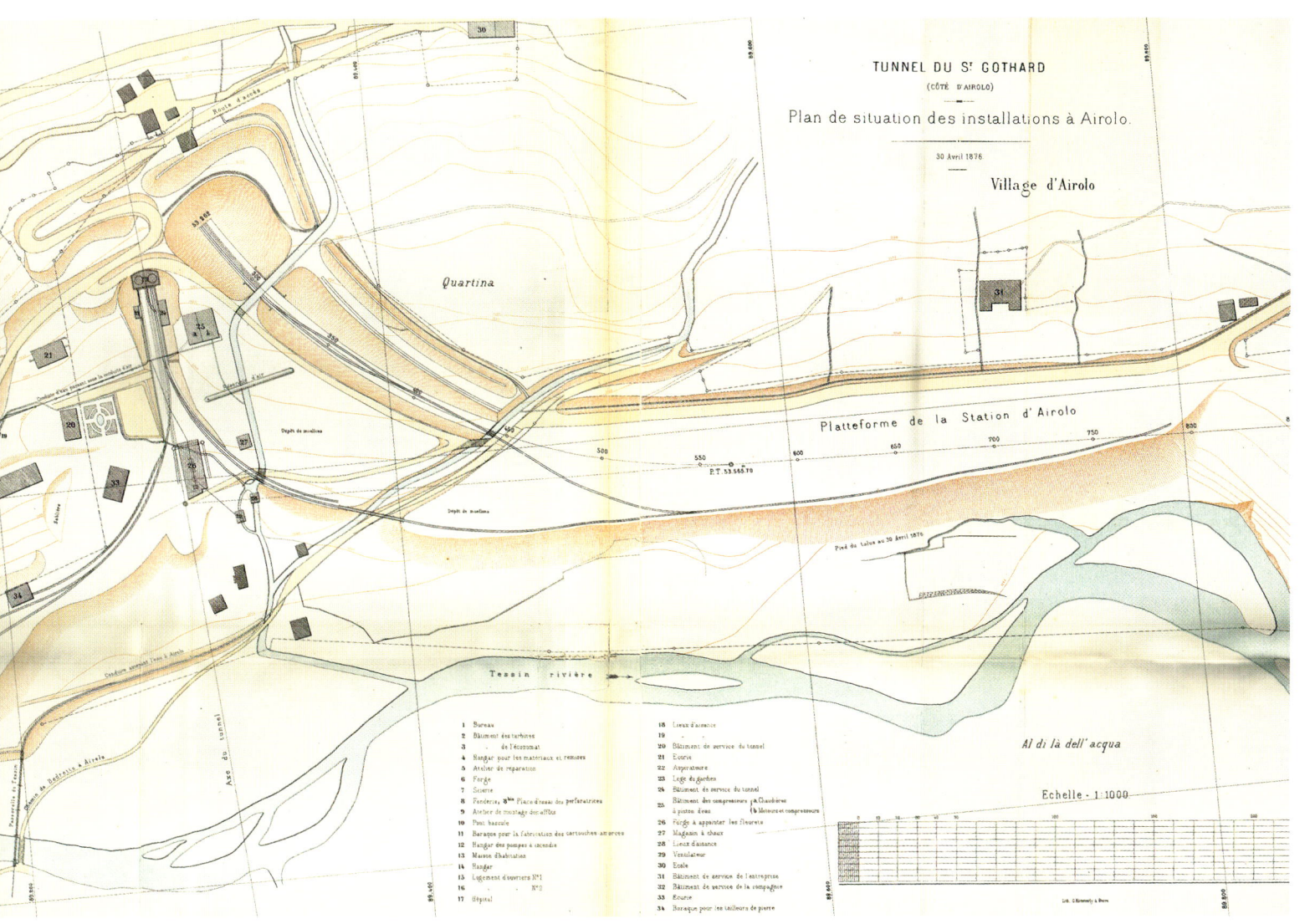

TUNNEL DU St GOTHARD

(CÔTÉ D'AIROLO)

Plan de situation des installations à Airolo.

30 Avril 1876

Village d'Airolo

Quartina

Platteforme de la Station d'Airolo

Tessin rivière

Al di là dell' acqua

Echelle - 1:1000

1   Bureau
2   Bâtiment des turbines
3   —   de l'épuisement
4   Hangar pour les matériaux et remues
5   Atelier de réparation
6   Forge
7   Scierie
8   Fenderie, 8bis Place d'essai des perforatrices
9   Atelier de montage des affûts
10 Pont bascule
11 Baraque pour la fabrication des cartouches amorces
12 Hangar des pompes à incendie
13 Maison d'habitation
14 Hangar
15 Logement d'ouvriers N°1
16 —   N°2
17 Hôpital
18 Lieux d'aisance
19
20 Bâtiment de service du tunnel
21 Ecurie
22 Arpenteure
23 Loge du gardien
24 Bâtiment de service du tunnel
25 Bâtiment des compresseurs a Chaudières
                              à piston d'eau b Moteurs et compresseurs
26 Forge à appointer les fleurets
27 Magasin à chaux
28 Lieux d'aisance
29 Ventilateur
30 Ecole
31 Bâtiment de service de l'entreprise
32 Bâtiment de service de la compagnie
33 Ecurie
34 Baraque pour les tailleurs de pierre

Construction Management and Construction Progress

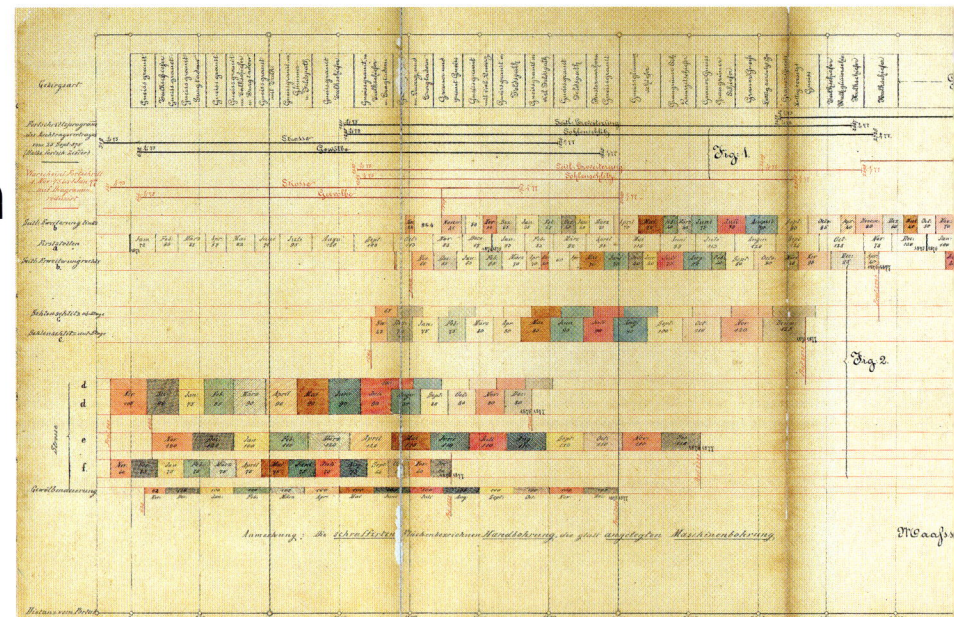

▷ "Graphical representation of the probable progress in construction on the Göschenen side, 1st November 1875 to 31st December 1876". Above are shown horizontally the rock layers to be driven through up to a distance of 4'200 m from the tunnel portal, left vertically the parts of the excavation (top heading, bottom heading, benches, etc.).

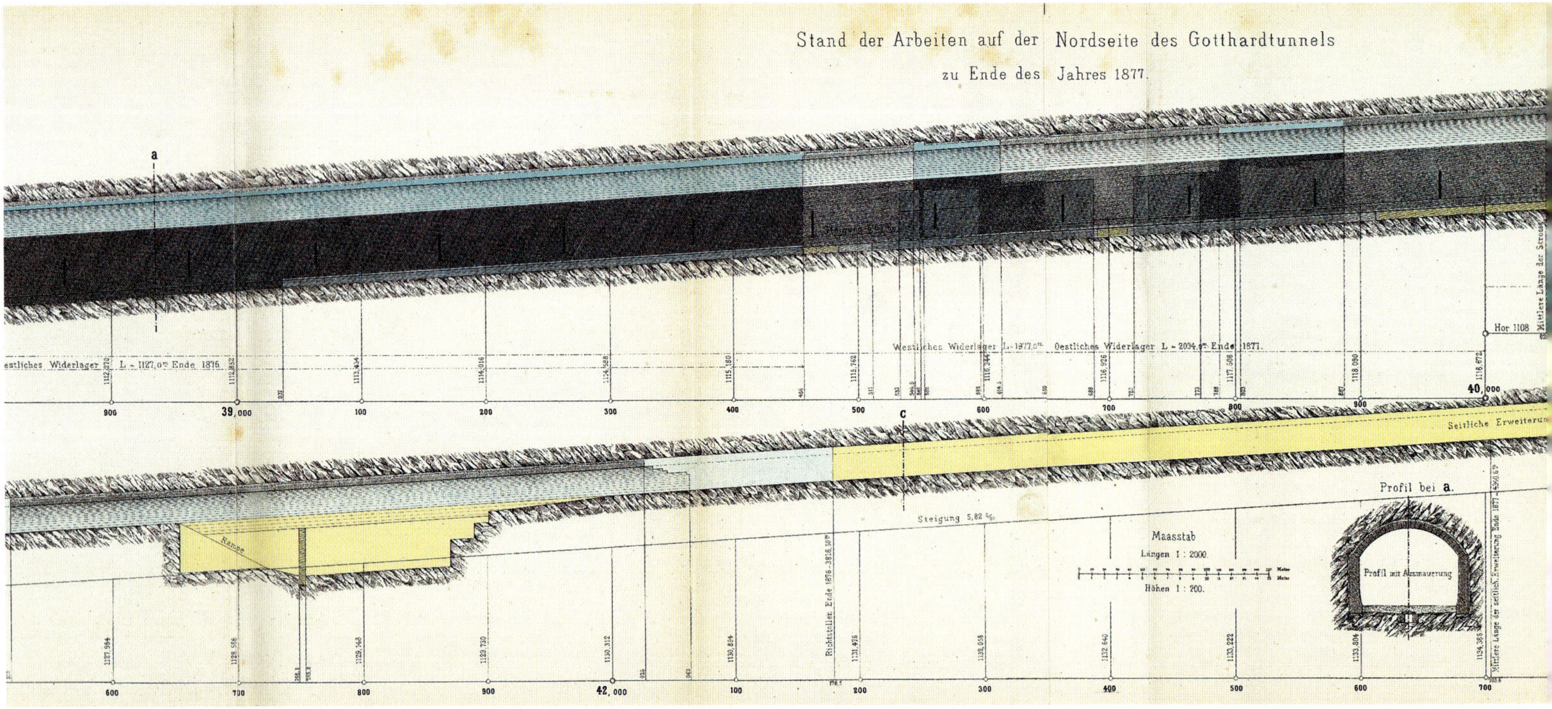

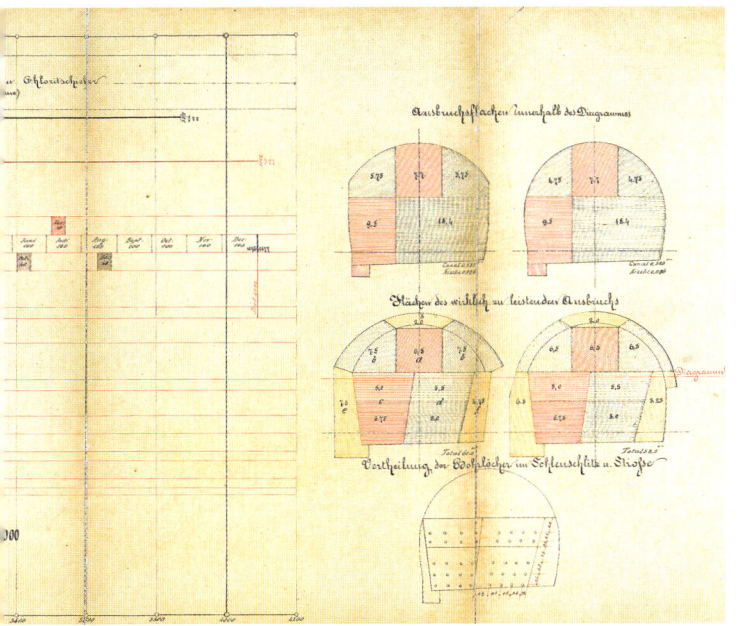

△ Construction train leaving the tunnel in Göschenen

◁ Progress made on the north side, December 1877

Construction Methods

Appendices to the contract with Louis Favre: Tunnel profiles

◁ Profile with roof arch (segmental arch)
◁▽ Profile with full masonry circular arch, arch consisting of hewn stones of equal height

▽ Circular masonry arch with invert
▽▽ Full masonry arch with drainage in invert

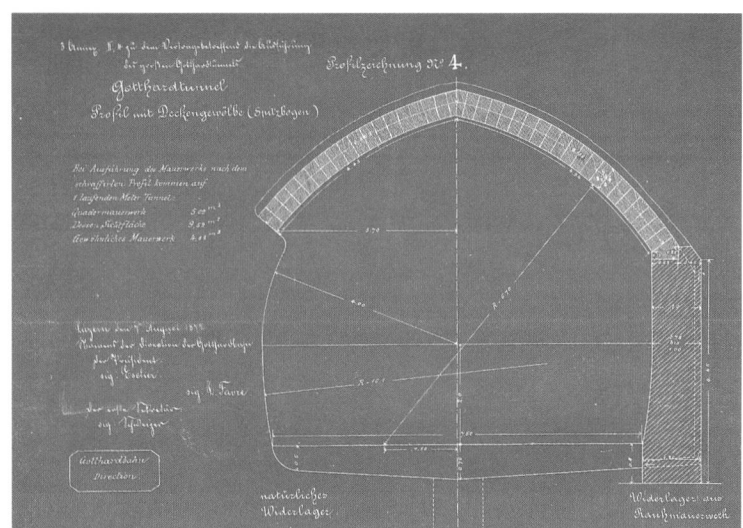

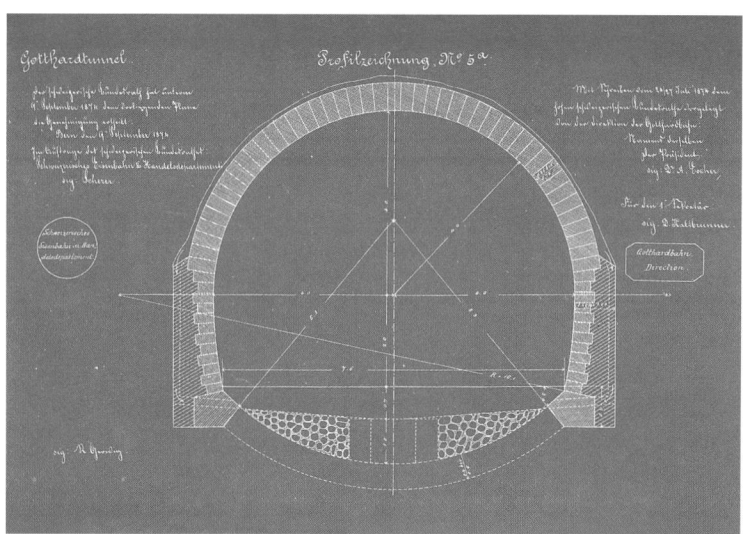

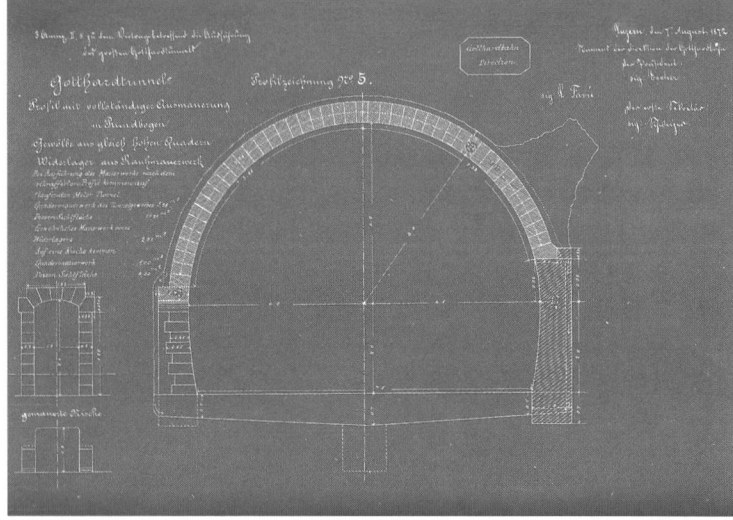

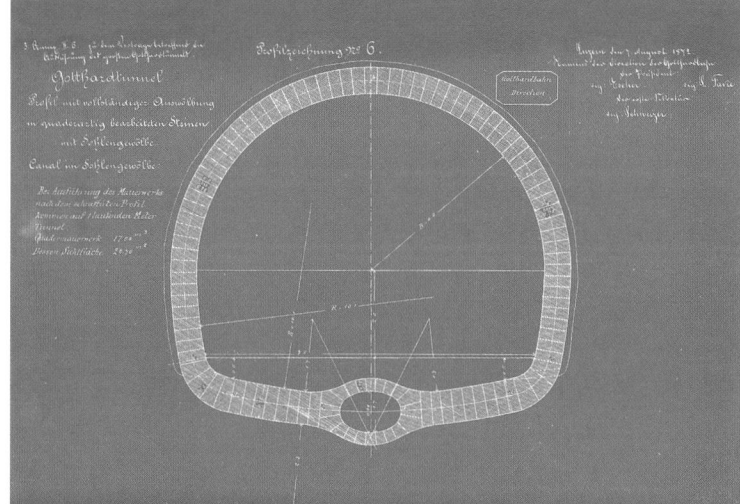

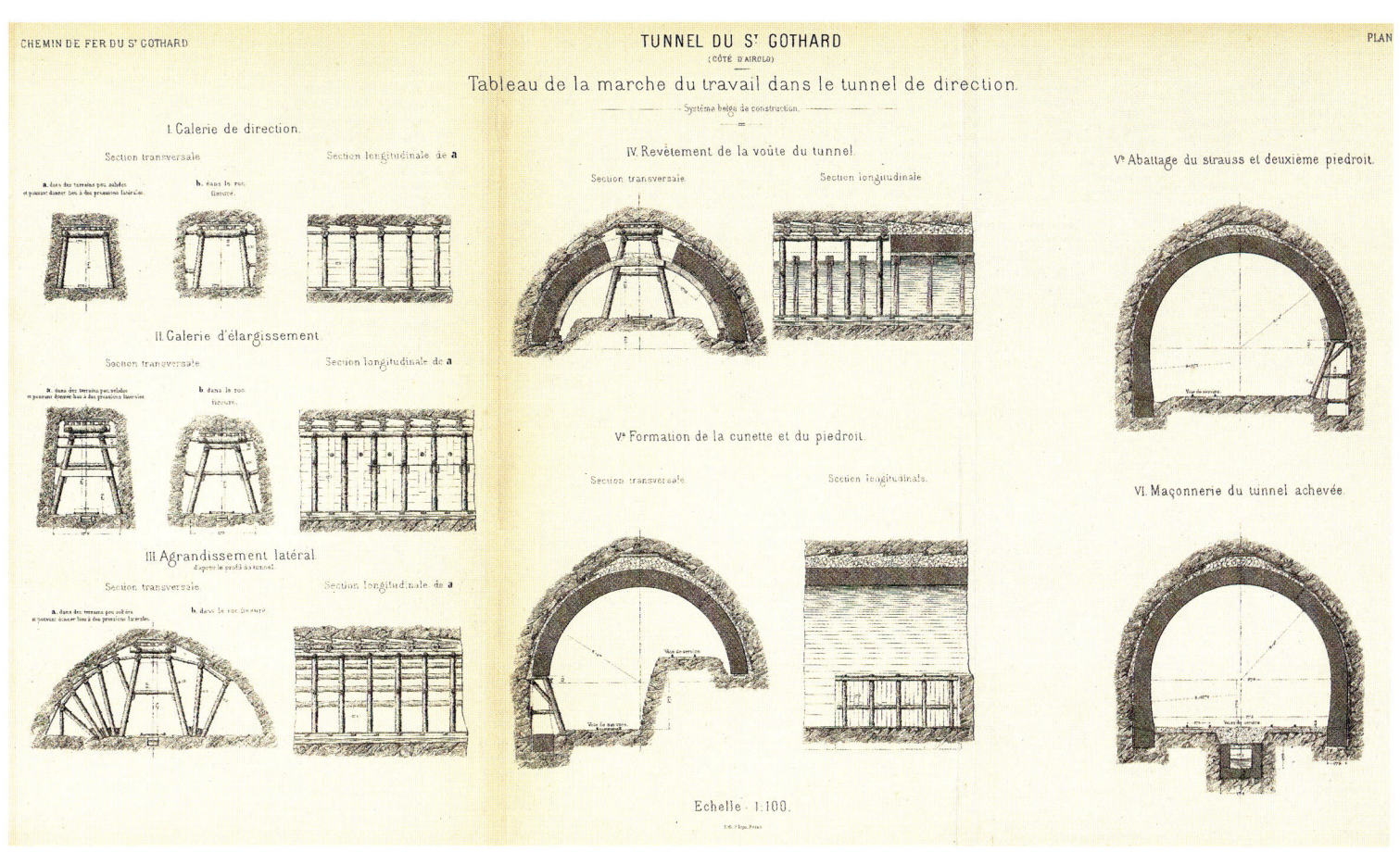

△ Construction stages

Tunnel Drilling Machines

△ Different drilling machines used in the construction of the Gotthard Tunnel. Systems (from top to bottom) Sommeiller, Ferroux, unknown, Turrettini and Colladon, McKean, Dubois and François. The drilling machines were operated by compressed air.

△ Compressed air operated engine for pulling the trains for the construction work

▷ Drilling rig with drilling machines. The System Ferroux is shown in the picture. At the beginning the installation material and the drilling machines were taken over from the Mont-Cenis-Tunnel (1857–1871).

▷ Tunnel drilling machine in operation.
Wood engraving

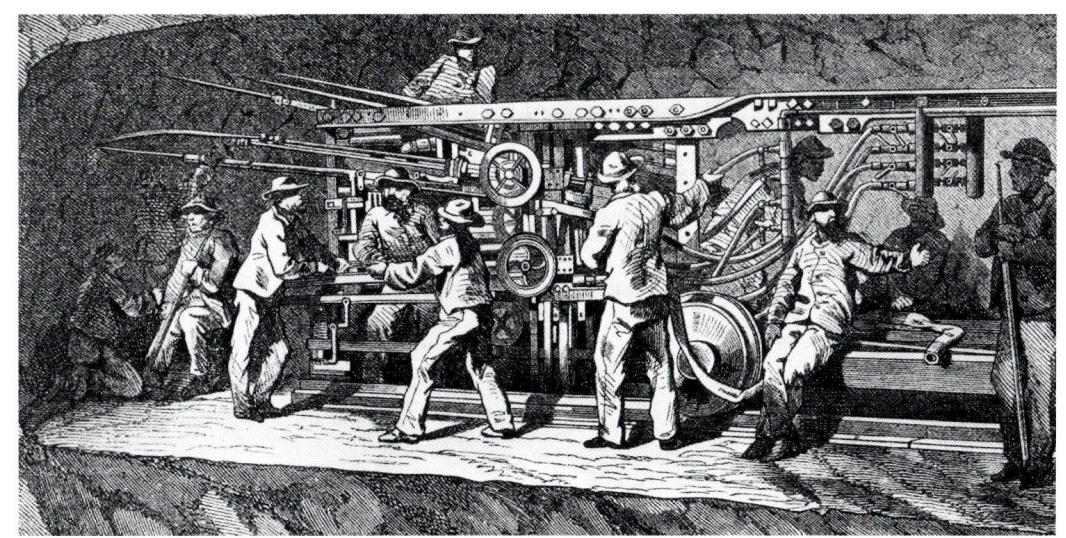

▽ Tunnel drilling machine System McKean

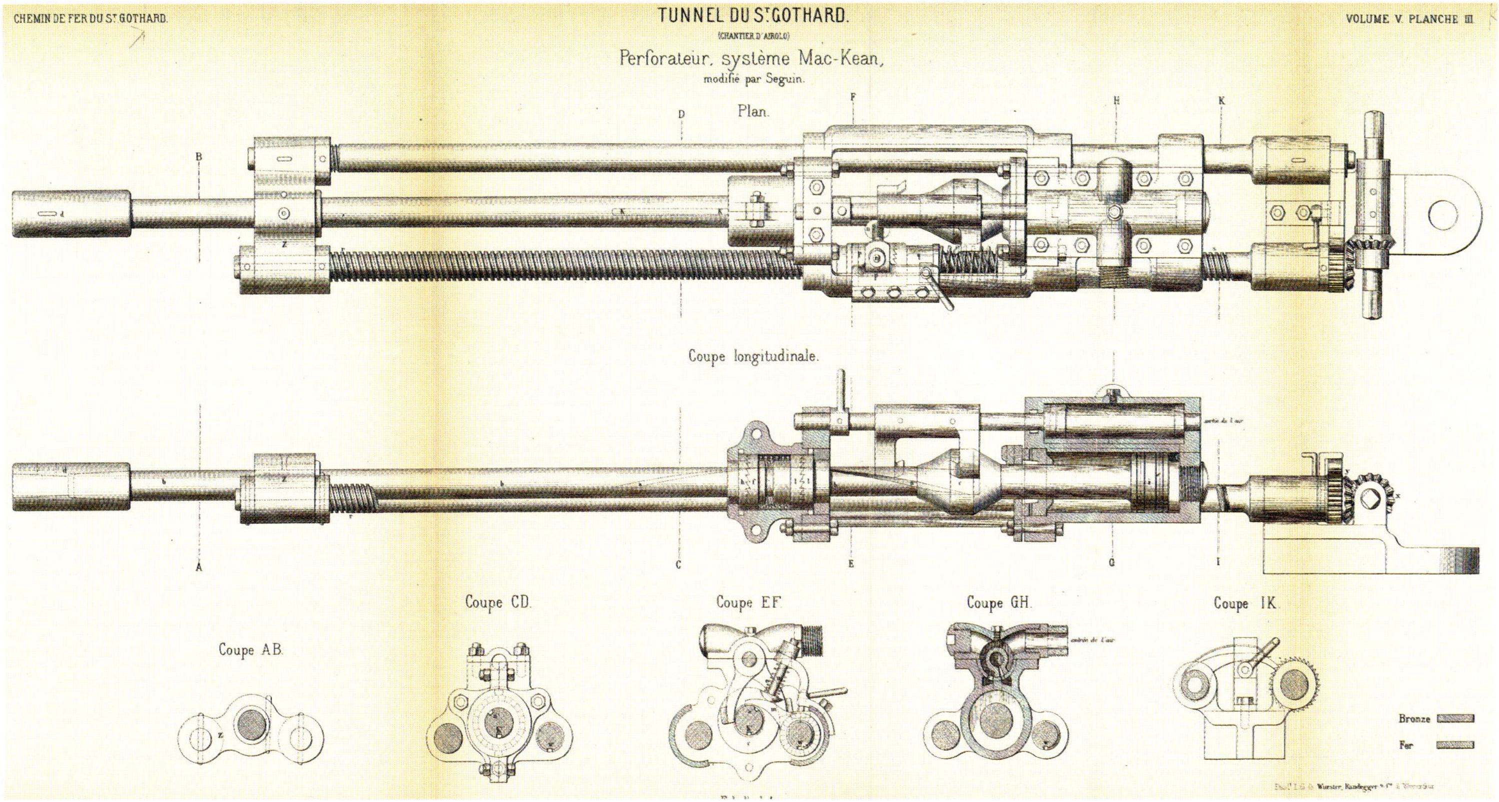
CHEMIN DE FER DU St GOTHARD.
TUNNEL DU St GOTHARD.
(CHANTIER D'AIROLO)
VOLUME V. PLANCHE III
Perforateur, système Mac-Kean,
modifié par Seguin.
Plan.
Coupe longitudinale.
Coupe AB.
Coupe CD.
Coupe EF.
Coupe GH.
Coupe IK.
Bronze
Fer

The Tunnel Breakthrough on the 28ᵗʰ February 1880

"Unexpectedly a slab of rock breaks away, a second follows, there is a noise – a jerking iron rod breaks through. The workers step back in a hurry. What is it? Then they suddenly realise: it is the other crew, who from Airolo on the south side with the trial boring tool, with the iron rod – iron is stronger than rock – the Gotthard has been driven through!

They are full of emotion, they cry out, they dance – they rush towards the iron tool, everyone wants to grasp it, but they burn their hands in the process. The iron is red hot. They have to let it keep on going. It doesn't matter, they don't feel any pain, they are full of jubilation. That piece of iron is no longer here, but the tunnel, this hole in the rock, is still here!

Watch out, they call from the other side. Attenzione, viene il signor Favre. What do you think? Favre is already dead. Povero padrone. But despite that should he come? A metal box is pushed through the hole. Open it, they cry from the other side. They open it. It contains a picture, a picture of Favre. Engineer Lusser sent it. Favre should be the first one to go through the tunnel, from south to north. (Moeschlin, 1957)

△ Scenes of triumph after the last blasting operation. Chief engineers Rossi and Stockalper embrace each other. Wood engraving based on a drawing by Jauslin

▷ Telegram from engineer Zollinger to the Gotthard Railway Management.

"Lucerne, 28.2.1880, 9 p.m.: Gotthard Railway Management. The first drilling machine has just broken through the separating rock wall. We are joined to Airolo. Göschenen is celebrating. Zollinger!"

△ A telegram from Engineer Zollinger to Engineer-in-Charge Bridel.

"Lucerne, 29th February 1880 in the morning: Engineer-in-Charge Bridel, Lucerne. Yesterday evening at six forty five the trial boring tool cut through the separating wall of rock, and so the difference between the borings is practically zero. Thus at five thirty this morning going in to begin the last drilling operation. Towards midday the breakthrough was completed. Presentation of medals on Monday morning!"

▷ Telegram of the Ticino government to the Gotthard Railway Management.

"Locarno, 1st March 1880, 4.05 a.m.: Gotthard Railway Company, Lucerne. Congratulations, wonderful, best wishes on the breaktrough of the big tunnel! Government of Ticino."

Construction Procedure – Costs – Personnel

"They work in three shifts. Each day, sometimes in the light, sometimes in the dark, they march into the tunnel. They follow the service rail track, they stumble over the railway sleepers, until they come to the rolling trucks and still further to the drilling rig and the drill carriage, the jack leg with the six drilling machines. And now the 15 workers are in the narrow 2.4 m wide and 2.5 m high openings, which is where they work. To the left and right of the drilling rig they have 50 to 60 cm in which to move around. They check that the rig is securely wedged. They turn the screw rods, until the back ends of the six drilling machines are set. After they have checked that the air tubes are correctly attached to the air pipes, they open the valves on the rig and let the compressed air into the drilling machines, which all now beat against the working face together. From time to time they change the blunted drilling bits.

When they have completed the 24 holes they drive the rig with the reservoir wagon and the drill carriages 150 metres back into a siding (because the pieces of rock can fly back that far),

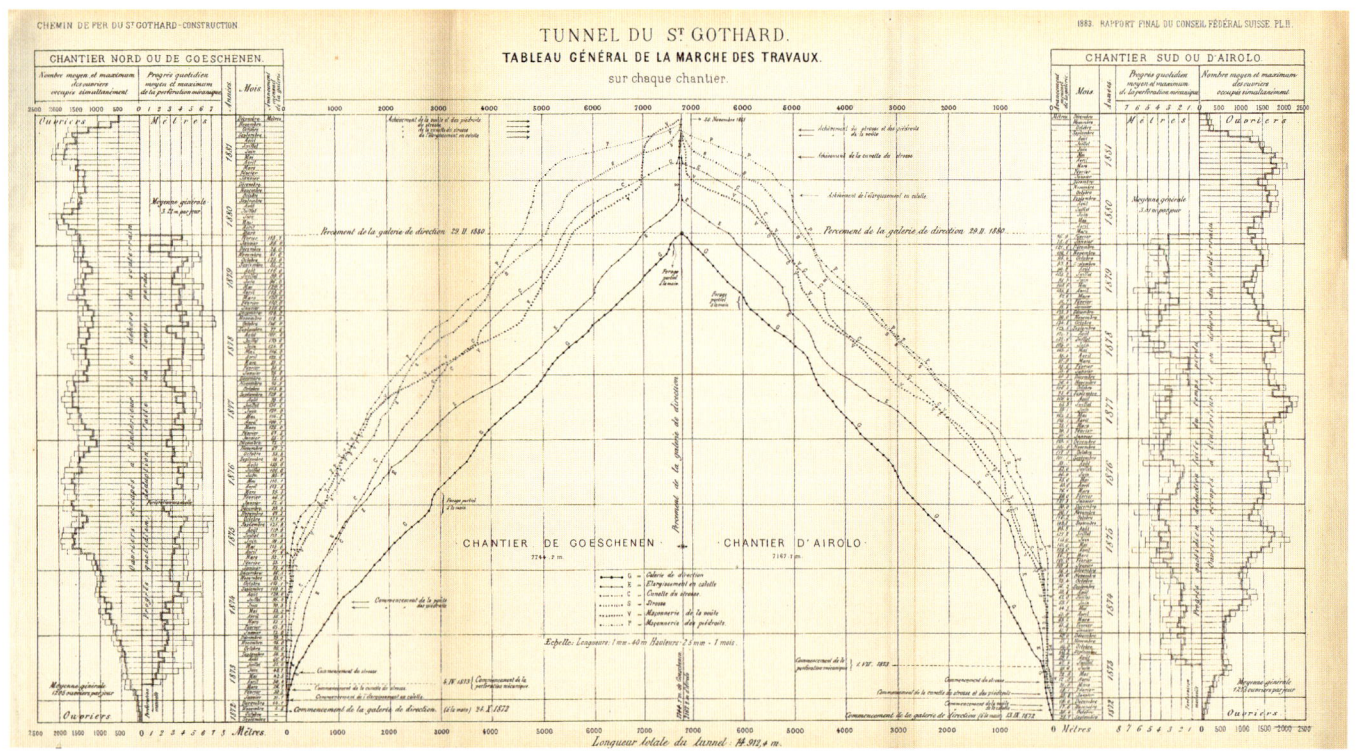

◁ Construction progress at the Airolo and Göschenen construction sites

they replace an old worn machine, during which time the blasting crew insert the sticks of dynamite into the middle six holes…
And thus the round goes on and on: forwards, drilling, backwards, filling the holes with dynamite, detonating, mucking, forwards, drilling again, day and night."
(Moeschlin, 1957)

Mois	Années										
	1872	1873	1874	1875	1876	1877	1878	1879	1880	1881	1882
Grand tunnel du St-Gothard (14.984 km.)											
Janvier . .	—	335	1215	2162	3079	2774	3072	2646	2718	2952	—
Février . .	—	403	1172	2236	3094	2797	2908	2663	2814	3093	—
Mars . . .	—	617	1372	2357	3086	2969	2822	2813	2730	3404	—
Avril . . .	—	650	1486	2937	3070	3231	2901	2539*	3013	3230	—
Mai . . .	—	947	1819	3329	3397	3381	3135	2486	3146	3628	—
Juin . . .	40	1036	1774	3350	3253	3626	3236	2639	3405	3319	—
Juillet . .	77	945	2093	3466	3344	3874	3022	2630	3330	3213	—
Août . . .	75	952	2131	3223	3423	3818	3187	2539	3199	2782	—
Septembre .	96	935	1992	2891	3153	3412	3040	2783	3051	2309	—
Octobre . .	126	1024	1971	2517	2940	3498	2823	2968	3162	1462	—
Novembre .	229	1092	1887	2697	2911	3305	2589	2875	2937	895	—
Décembre .	272	1149	1962	2947	2650	2984	2552	2756	2781	678	—
Moyenne	*181*	*840*	*1740*	*2843*	*3117*	*3305*	*2940*	*2695*	*3024*	*2580*	—

Chapitres des comptes de la Compagnie.	Grand tunnel. Longueur : 14.984 km.			
	Devis de mars 1879. Justification financière.	Dépenses effectives fin 1885.	Différence avec le devis.	Dépense moyenne par kilomètre.
	Fr.	Fr.	Fr.	Fr.
I. Remboursement des dépenses antérieures à la constitution de la Compagnie . .	162,200	162,200	—	10,800
II. Frais de formation du capital . . .	—	—	—	—
III. Administration centrale (non compris la direction technique des travaux) . .	1.010,000	874,894	— 135,106	58.400
IV. Intérêts des actions et obligations pendant la période de construction . . .	2,657,161	2,581,400	— 75,761	172,300
V. Construction *	60.694,100	63.048,087	+ 2,353,987	4,207,700
Ensemble	64,523,461	66,666,581	+ 2,143,120	4,449,200
A déduire : Recettes nettes des lignes tessinoises subalpines pendant la période de construction . .	—	—	—	—
Dépenses totales	64,523,461	66,666,581	+ 2,143,120	4,449,200

The Gotthard Railway Line after its Opening in 1882

"Compared to the 14 year construction period of the Mont-Cenis Tunnel (12 km) that for the Gotthard Tunnel (15 km) at just over 9 years was very favourable. Without the discovery of the drilling machine by the Geneva physicist Colladon, which was already employed at Mont-Cenis, and the continual improvement of the construction systems it would never have been possible.

On the 1st January 1882 the operation of the railway through the great tunnel could be started provisionally and with a single track. At the end of spring the two access ramps were completed. The opening ceremonies in Lucerne and Milan lasted from the 22nd to the 25th of May. Alfred Escher was also invited, but could not participate because of a serious illness. On

△ Arrival of the opening ceremony train on the 23rd May in Bellinzona

◁ Airolo at the time of the tunnel opening. Embankment constructed with debris from the tunnel

the 6th December he died in his home in Zurich. More than anyone else through his bold plans and tireless efforts he gave decisive impulses to the economic development of Switzerland and his home city Zurich in the 19th century.

On the 1st June 1882 the Gotthard Railway Line started normal time-table operations. Its construction took a toll of 307 workers lives and up to the start of operations it swallowed up 227 million Swiss Francs. After just a few years the traffic performance of the line exceeded all expectations.

By 1904 it was possible to extend the major part of the stretch between Goldau and Chiasso to double track."
(Wyss-Niederer, 1979)

△ Göschenen after the opening of the tunnel. View to the south

▷ The railway station area at Bellinzona with the new station building in 1882

The Gotthard Railway Line after its Opening in 1882

◁ Wassen, view to the north. Left in the background the upper Maienreuss Bridge

◁ Amsteg, in the background right the Kerstelenbach Bridge

▷ Railway and road in the Biaschina around 1882. The step in the valley floor above Giornico was overcome by two loop tunnels. Solutions were also considered involving cable or cog railway! Above is the Pianotondo viaduct

▽ Engine for the first mail trains through the tunnel from the 1ˢᵗ January 1882

▽ Engine for the opening ceremony, 1ˢᵗ June 1882

The Simplon Tunnel

	Simplon I with parallel gallery	Simplon II (enlargement of the parallel gallery)
Total length	19'803 m	19'823 m
Construction period	8 years	9 years
Opening date	1st June 1906	4th December 1921
Total costs	78 million SFr.	34 million SFr.

△ The horse-drawn mail coach still runs over the Simplon Pass (Simplon road at the south portal, right Tunnel I, left entrance to pilot gallery)

"But it was possible"

"Already in the year 1886 feasible plans for a Simplon Tunnel were prepared. The corresponding studies are very promising but the interested cantons did not want to commit themselves. About thirty projects with detailed plans, some of which are not practicable, pass from one draw to another. Not till many years later and after all kinds of experience had been gained at Mont-Cenis and Gotthard, was it finally decided to go ahead with the Simplon Tunnel, and moreover a base tunnel, despite the 'big uncertainty' one would encounter in the depths of the tunnel: the high temperature. According to observation this increases 1 °C for every 52 m into the mountain, which posed problems for the ventilation and cooling. One had already had 63 workers dying of suffocation at the Hauenstein tunnel. With the Simplon this dare not be repeated.

When on the 18th June 1878 the last stretch of the railway was opened to traffic in Brig it was natural that at the banquet the main point of discussion was the possibility of a Simplon tunnel.

The council of Canton Vaud wrote at the time:'...cette enterprise gigantesque est bien peu probable' (this gigantic construction work is hardly possible).

But it was possible!

In 1857 Italy had boldly started to drive a tunnel through the Mont-Cenis. At the beginning drilling was done by hand; here for the first time on the European mainland the technique of compressed air drilling was used, an invention of the Milan engineer Giovanni Piatti. Piatti registered his patent in Liverpool, where a tunnel was being built. This invention was contended in his native Italy, and it was not till much later that he was rehabilitated. On the 17th September 1871 the Mont-Cenis Tunnel, with its 12.849 km was opened to traffic. Two years later the driving operations were begun on the St. Gotthard Tunnel.

In the meantime something happened that was to benefit all drilling work in rock: the Swedish chemical engineer Alfred Nobel registered a patent for the use of nitro-glycerine as an explosive material. The first to profit from this was the Gotthard Tunnel, and then later the Simplon Tunnel.

After numerous projects and scientific studies and pamphlets, hundreds of meetings with experts from all fields of civil engineering it was decided to go ahead with the construction of the railway tunnel through the Simplon. On January 1st 1890 the Jura-Simplon Tunnel Railway Company was constituted. On the 23rd September 1893 the contractor Brandt, Brandau and Co. was awarded the contract to build the Simplon Tunnel. In Berne on the 25th November 1893 the Simplon Construction Contract was signed between Switzerland and Italy and on the 22nd February 1896 the Simplon Convention was signed in Rome." (Steiner-Ferrarini, 1992)

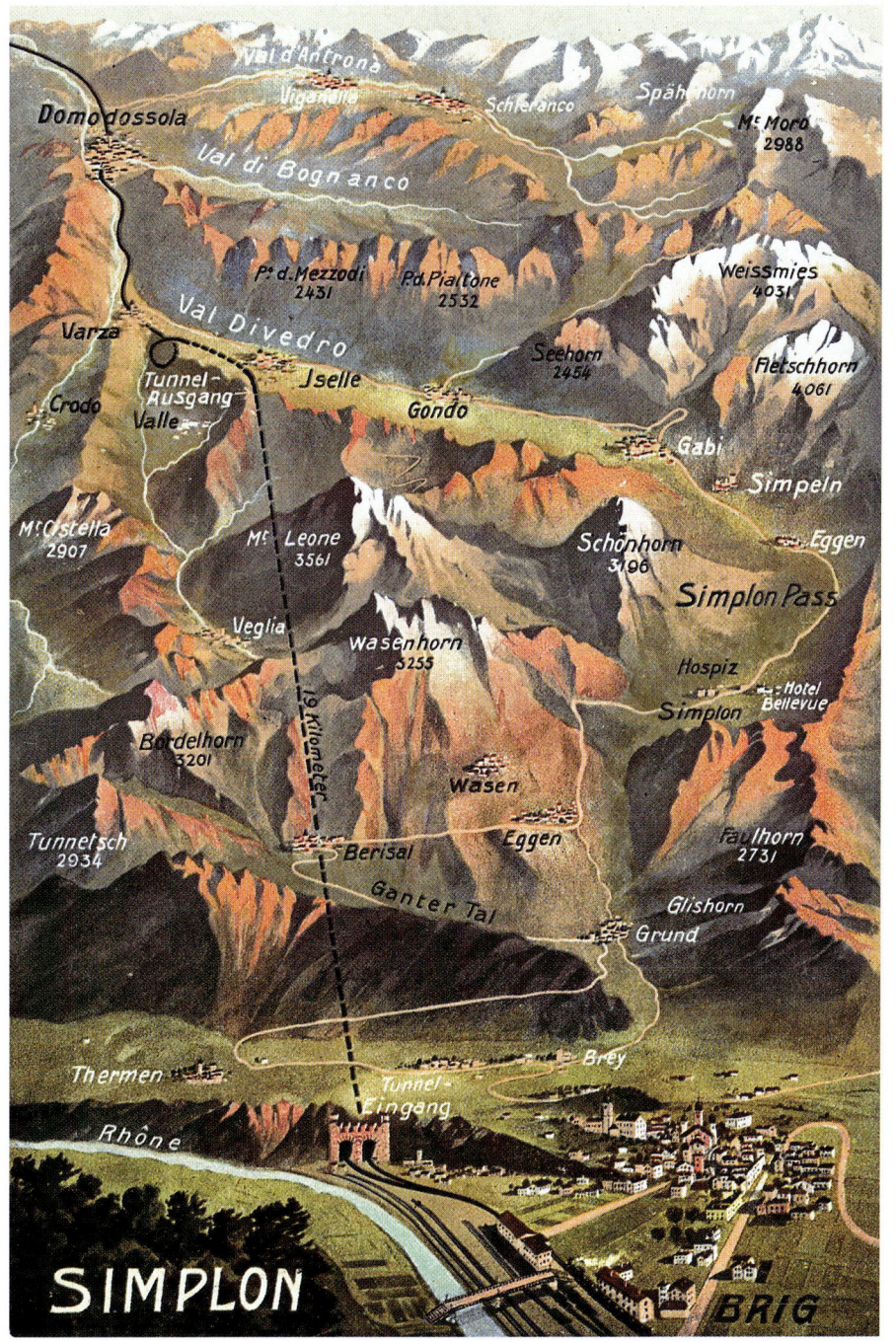

▷ Postcard with the Simplon Tunnel

Geological Predictions

"Member of the Council of the Swiss Federation Ed. Sulzer of Winterthur gave a lecture on the 'Construction of the Simplon Tunnel' at a banquet of the Natural Science Foundation in St. Gall on the 26th January 1904 and later on other occasions, especially at the annual meeting of the Swiss Society for Scientific Research in Winterthur. In it he claimed that the geo-logical predictions on what one would encounter in the tunnel were completely wrong and continually he intimated that the contractor, deceived by the wrong opinions of the geologist, ventured to construct the tunnel and thereby had now got into difficulties. His assertions of course found their way into all the newspapers are were even echoed in the ne-gotiations with the authorities, and a mockery was made of geology and ridicule poured upon the geologists. This was much easier, as Mr. Sulzer himself indulged in disparaging but very general remarks, mentioning no names and still less quoting from records, in which the geologists were purported to have made such blatant mistakes."

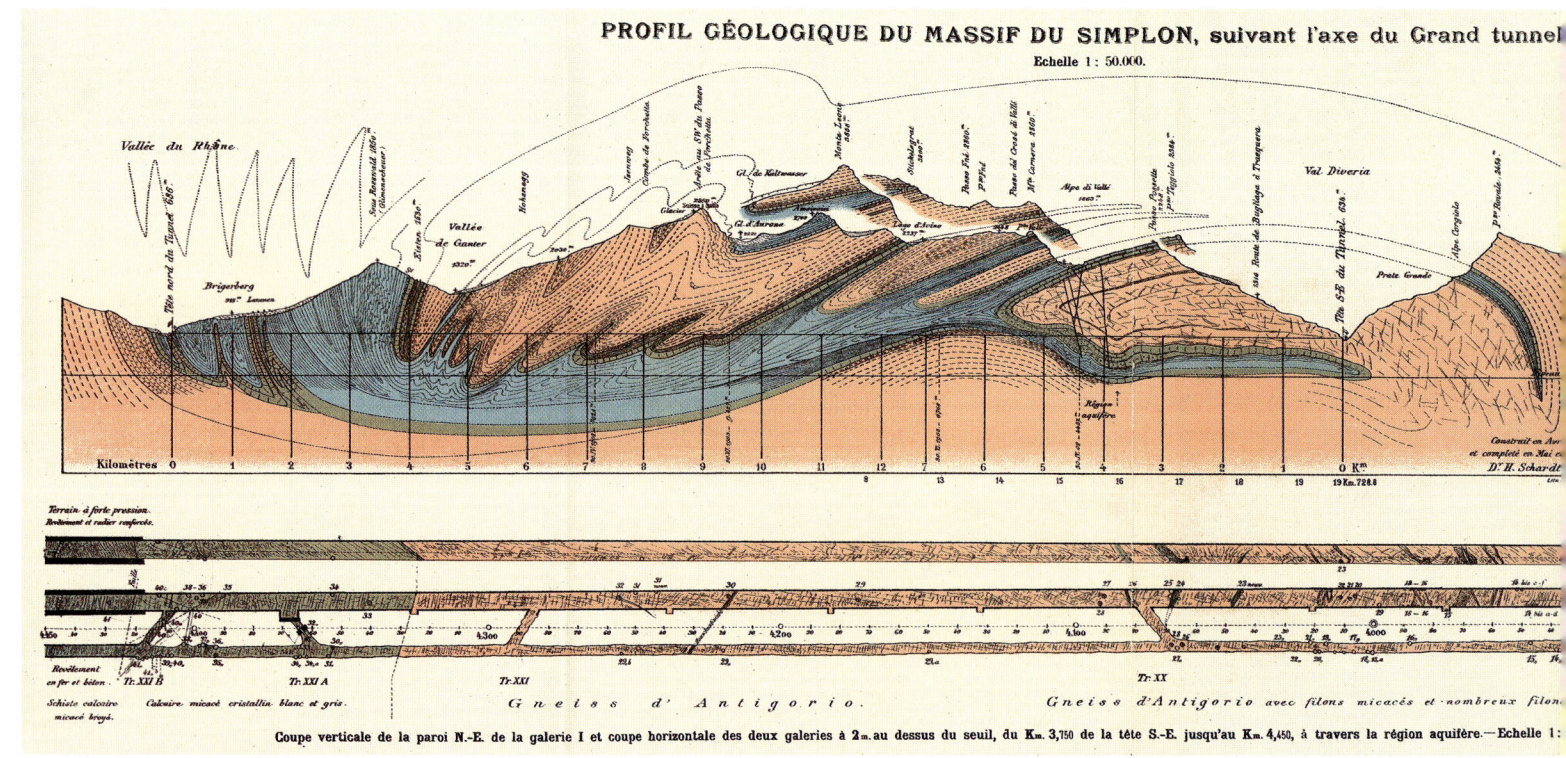

PROFIL GÉOLOGIQUE DU MASSIF DU SIMPLON, suivant l'axe du Grand tunnel
Echelle 1 : 50.000.

Coupe verticale de la paroi N.-E. de la galerie I et coupe horizontale des deux galeries à 2 m. au dessus du seuil, du Km. 3,750 de la tête S.-E. jusqu'au Km. 4,450, à travers la région aquifère. — Echelle 1:

(Quotation from "On the geological predictions for the Simplon Tunnel; Answer to the attacks of the Council Member Mr. Sulzer, written by Prof. Dr. Albert Heim, on behalf of the Simplon Commission dealing with the geology."
Eclog. Geol. Helv., 1904)

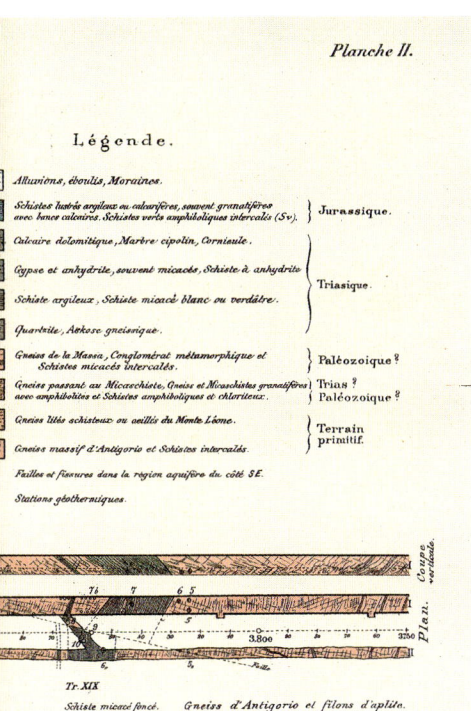

◁ Geological profile of the Simplon massif (along the tunnel axis)

△ The first projects for driving through the Simplon emerged in 1853. With the further development of the tunnel projects again and again new geological predictions were made. (The figures come from the publication of Prof. H. Schardt "The Scientific Findings of the Simplon Tunnel", lecture given at the 87th annual meeting of the Swiss Society for Scientific Research in Winterthur, 30th July to 2nd August 1904.)

Construction Company

"As a result the construction company for the Simplon Tunnel, Brandt, Brandau & Co., with headquarters in Winterthur was formed. This company, which comprised Engineer Alfred Brandt from Hamburg, Engineer Karl Brandau from Kassel and the firms Locher & Co., Zurich, Sulzer Brothers in Winterthur and the Bank in Winterthur, offered to carry out the construction of the Simplon Tunnel. Thereby they proposed a completely new, previously unknown method of construction. Whereas all earlier projects involved the construction of a single double track tunnel tube, Brandt, Brandau & Co. wanted to build two single track tubes, whose bases were at the same elevation and axes were 17 m apart running parallel to each other. It was planned to drive the bottom headings simultaneously and to connect them every 200 m by a cross tunnel. Provisionally only the one tube should be completed as a tunnel and one should wait with the second one until the traffic conditions necessitated it."
(Vom Bau des Simplontunnels, 1921)

Nationalrat Dr. Eduard Sulzer-Ziegler   Oberst Dr. Eduard Locher   Chefingenieur Karl Brandau
Oberingenieur Dr. Hugo v. Kager   Chefingenieur Alfred Brandt †   Oberingenieur Dr. Hermann Häussler
Oberingenieur Hans Beissner   Oberingenieur Dr. Konrad Pressel

Les Ingénieurs principaux des travaux du Simplon.   Die Erbauer des Simplon-Tunnels.
I principali Ingegneri dei lavori del Sempione.   The Chief-Engineers of the Simplon Tunnel.

▽ Installations for excavating the tunnel

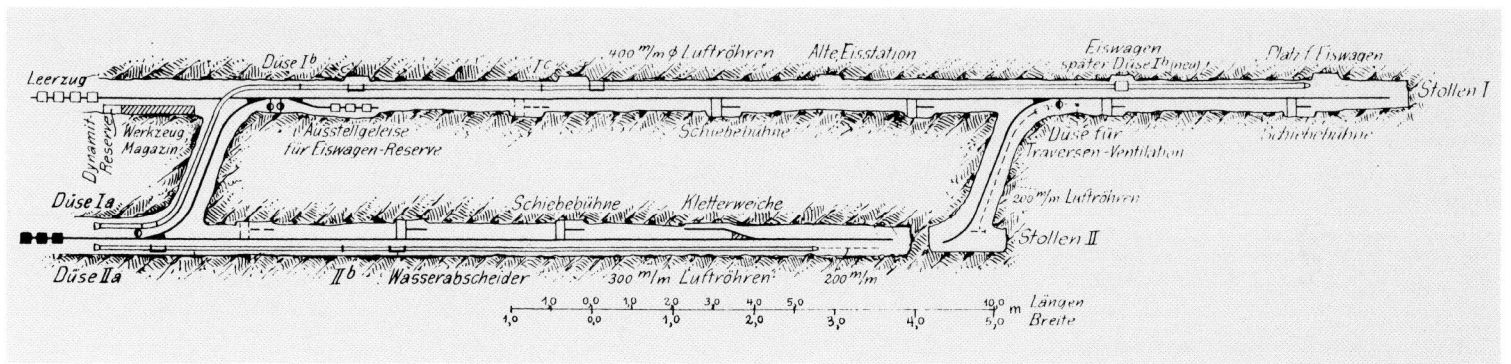

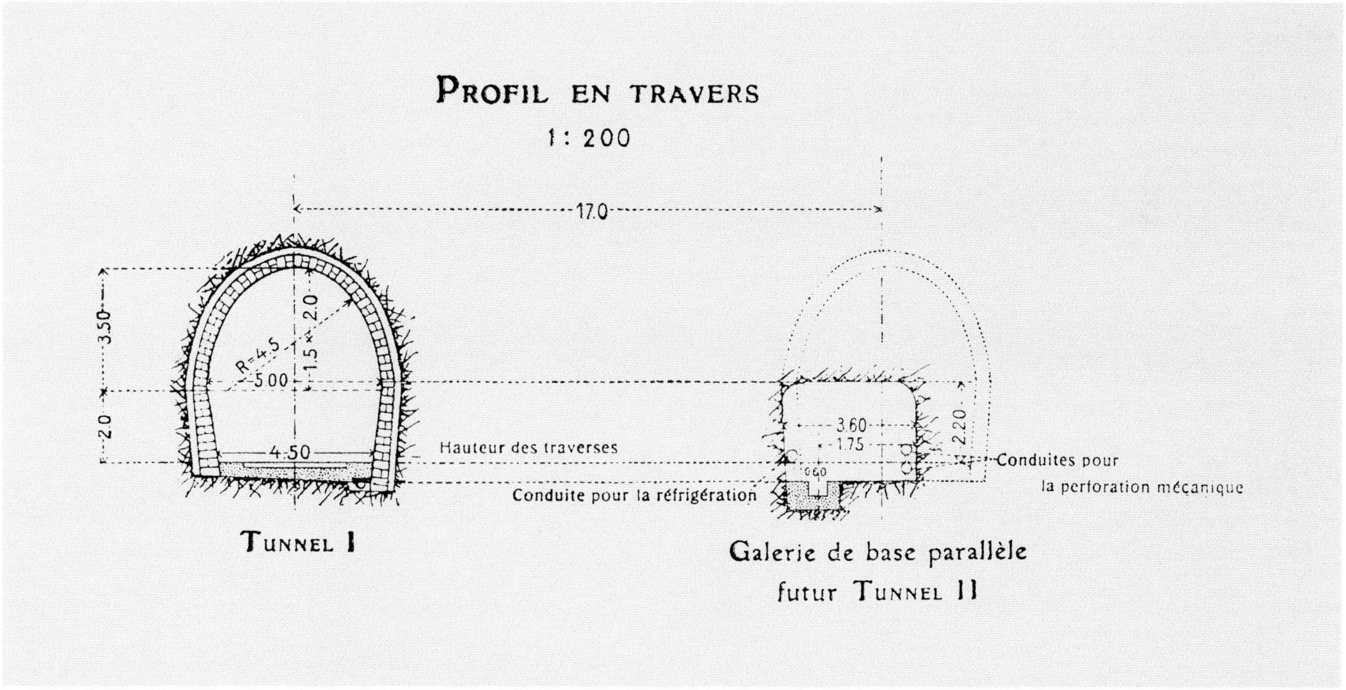

PROFIL EN TRAVERS

1 : 200

TUNNEL I

Hauteur des traverses

Conduite pour la réfrigération

Galerie de base parallèle
futur TUNNEL II

Conduites pour
la perforation mécanique

Production of Energy

"The power needed to run the machines was taken from the Rhone, which was dammed 4 km above Mörel. The water was fed into a reinforced concrete closed channel of square section, which for long stretches was carried on reinforced concrete supports along the slope. The surge tank, a reservoir from which the 1.60 m diameter steel penstock carrying the water to the power house, is situated at the end of this channel, above the mouth of the gorge carrying the Massa stream coming from the lower Aletsch glacier. The head difference of this penstock amounts to 52 m and the performance corresponding to the water power produced was more than 2200 horse power."

(Rosenmund, 1905)

◁ Collection of the Rhone waters for the hydroelectric power plant below Mörel

▷ Reinforced concrete channel of the Rhone hydroelectric power plant

▷ Inside of the dynamo building: production of the electrical energy for light and power transmission

Excavating of the Two Pilot Headings

Brig side of tunnel

"The rock driven through was generally favourable along the whole length of the tunnel because of the steep dip of the rock layers over long stretches, which may be clearly seen from the exposed longitudinal geological profile; but there was no shortage of stretches, however, in which the rock was very fragile and of a squeezing character, necessitating manual excavation and time-consuming installation of the lining."

Iselle side of tunnel

"Already in the region of the gneiss on the stretch excavated up to September 1901 (from the south portal to 4357 tunnel metre) the horizontal layering of the rock proved to be an obstacle. The effectiveness of the blasting was poor and it was not possible to mine deeply. An attack achieved on average no more than a metre progress per day. Despite gathering together all experience and employing all conceivable means and despite the excellent organisation of the excavation work it was not possible to get ahead of the construction programme."
(Pressel, 1906)

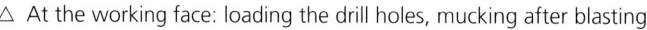
△ At the working face: loading the drill holes, mucking after blasting

◁△ Shell-like blocks breaking away at the working face and spalling in a top heading in Antigorio gneiss

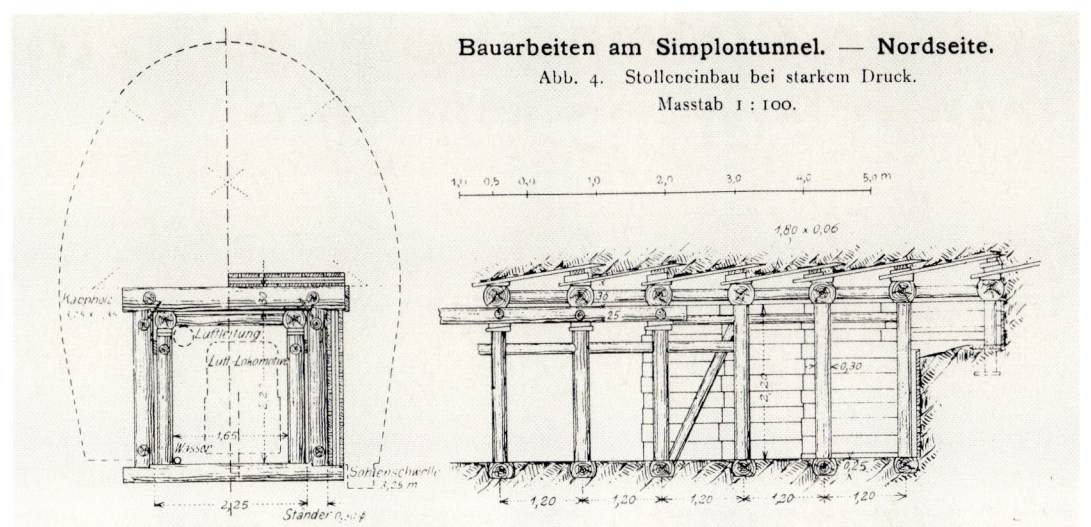

Bauarbeiten am Simplontunnel. — Nordseite.
Abb. 4.   Stolleneinbau bei starkem Druck.
Masstab 1 : 100.

△ Base heave leads to deformations in dewatering channel and causes flooding of the tunnel

▷ Deformed timbering in a cross-tunnel (south side)

Drilling Blasting Holes with the Brandt Drilling Machine

"For the drilling work on the Simplon Tunnel the Brandt drilling machine was employed. These are so-called rotary drilling machines in contrast to the percussion types, as were used in the Gotthard Tunnel. In the case of the latter the drill hole results from the blows of the drill on the rock just like with hand drilling.

The Brandt drilling machine works with water pressure. The pressure on the piston of a twin motor is transmitted by a cranks to a shaft, which causes a cylinder to rotate. At the same time this cylinder is subjected to a hydraulic pressure, whereby a shaft with the drill fixed to its front end is pressed against the rock. This drill is hollow and has three teeth on its rim. By pressing the teeth against the rock and the rotation of the drill the rock is loosened; the crushed material is washed away by a jet of water issuing from the inside of the drill.

Depending on the character of the rock, 9–12 drill holes of 9 cm diameter and up to 1.2 m long are driven into the rock face. Then the drilling machine placed on the jack leg is pulled back onto a rolling truck, the holes are filled with nitro-glycerine and provided with detonating fuses, which are ignited. The explosion of the charge causes a mass of rock to be separated from the rock face which should correspond to the depth of the drill holes."
(Rosenmund, 1905)

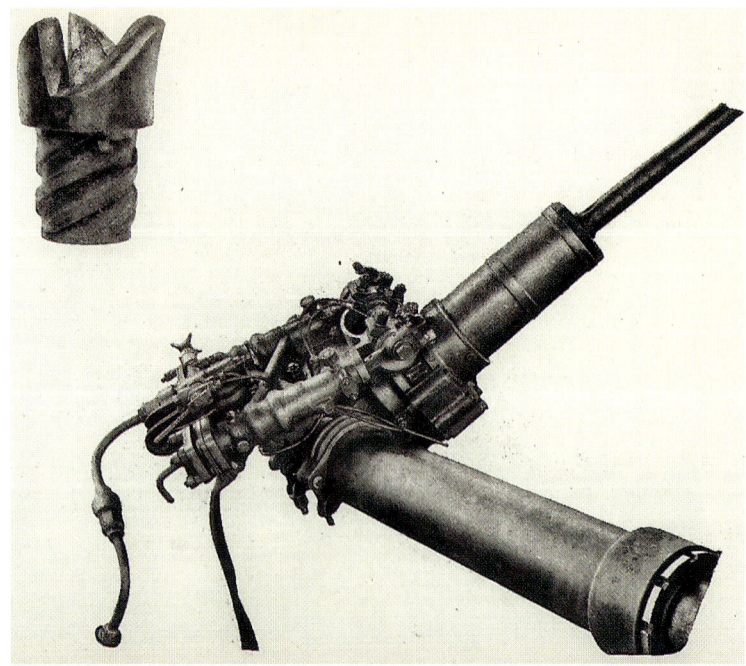

▷ As a rule three drilling machines are placed on a so-called jack leg and can be rotated about it both vertically and laterally. Thus three holes can be drilled simultaneously. During the drilling operation the jack leg can, by means of hydraulic pressure, be pressed against the side walls of the tunnel, whereby the machines are provided with the necessary support.

Production and Transport of Explosives

"The construction company is in principle bound by contract to purchase the explosives as indeed all nitro-glycerine products from a Swiss company manufacturing dynamite, which built a factory especially for this purpose in Gamsen (1884); with respect to other explosives, however, it was free to choose. Thus a Munich professor, Linde by name, who was famous for producing liquid air, offered his own discovery in the field of explosives to the Simplon Company. His product exhibited a higher explosive force than the gelatine-based explosive or was at least equivalent to it, and was at the same time much cheaper. Unfortunately, the reactions between liquid air and other bodies appear to give rise to a danger of explosion, so that its use in open excavations is for the present time unthinkable."
(Briger Anzeiger, 1899)

◁ Factory in Gamsen of the Federal Government's Explosives Company

△ Dynamite transport from Gamsen to the tunnel site with a horse-drawn carriage and police escort…

▷ …and by rail to the working face

▷ Production of nitrous acids as an intermediate product in the manufacture of dynamite

Enlarging the Tunnel to the Full Profile and Timbering

"From the bottom heading over lengths of approximately 100 m the excavation was taken upwards in shafts to the planned height of the tunnel roof and thus the top heading could then be excavated, working forwards and backwards parallel to the bottom heading. For each shaft there were two working positions and each forward-driven heading from the trailing shaft connected with the backwards-driven heading of the leading shaft. After completion of the two parallel top and bottom headings the full profile could be excavated piecewise and lined. The whole of the tunnel, even the parts in hard rock, was lined with masonry. In order to be able to carry out more easily the enlarging of the excavation and the lining of the tunnel the work proceeded at two levels, and extensive rock support was required. Therefore, especially in the parts of the tunnel where this enlargement to the full profile was being carried out, it was necessary to have timber supports, inside of which passing was rather restricted."
(Rosenmund, 1905)

▷ Enlargement of the tunnel roof

▷ Timbering in the tunnel roof

△ Vertical excavation upwards, bottom and top headings

△ Over large tunnel stretches where the roof area did not exhibit the squeezing phenomenon and thus no rock deformations were expected the timber support work was executed with a long trestle construction.

△ Enlargement work

Masonry Lining

On both the north and south sides of the tunnel the stone for the roof lining was obtained from nearby quarries operated by the company.

"Special regulations for the roof support:

1) ...

2) Both flanks of the masonry roof arch must always be constructed simultaneously and uniformly starting from the abutments. For large roof arches the falsework has to be loaded correspondingly at the crown, or the lining is to be constructed following the instructions of the resident engineer, in concentric rings or simultaneously at several places.

3) ...

4) After finishing the flanks from both sides the keystones have to be cut to exactly the right size and placed with care.

5) The actual arch lining begins at an inclination of the support faces of 4:1. Up to this height the lining is to be constructed as an abutment wall but nevertheless with radial support joints. In the case of large arches, however, already at 1.5 m above the springing line it is to be designed as an arch lining. If the proposal does not prescribe anything else the lining is to be constructed uniformly in its whole thickness in masonry."

(Besondere Bestimmungen der SBB, 1902)

Galleria: Allargamento   Tunnel: Elargiss

▷ Constructing the masonry lining for the walls

▷ The quarry on the north side was located at Massaboden, about 1.5 km from the tunnel portal

▷ Constructing the masonry lining

Transportation with the Tunnel Construction Railway

"On the completed tunnel section the transport took place using steam engines up to the two track marshalling area in the tunnel. From there to the pilot tunnel the trains are driven by compressed air engines. Every 24 hours there are three shifts, starting at 6 a.m., at 2 p.m. and at 10 p.m. At these times eight trains run one after another into the tunnel four to the excavation areas for the advanced bottom heading and four to where the complete face is being excavated and the masonry lining installed. Each of the first four trains is reserved for the transport of workers, using the special passenger trucks."
(Pestalozzi, 1904)

◁ The trucks in the headings were partly horse-drawn. We remember here for just one moment the vast number of horses that had to work underground. They all became blind and died in silent agony

▽ Exit of a train carrying workers on the north side, pulled by a special tunnel steam engine

▷ Electrically-operated crane for unloading the tunnel trucks

△ Crane for unloading and reloading in the tunnel

▷ Compressed air engine for operations at the working face made by the Swiss Railway Engine Works in Winterthur

Overcoming the Stretch with Squeezing Rock

"The tunnel stretch with the big cold springs had only just been overcome when the contractor was faced with even more difficult and time-consuming problems: They had come to a rock mass that exhibited large rock pressures in every direction. It was like a mass of dough, consisting along the axis of the tunnel – according to Prof. Dr. C. Schmidt of Basle – of micaceous limestone. All attempts to penetrate the rock using conventional mining techniques with a complete frame with larch timbers of diameter 50–60 cm side by side failed; the timber frames could not resist the pressure. Even oak timbers of cross-section 40×40 cm were crushed, although the dimensions of the frames were on average no more than 1.95 m wide and 2.0 m high."
(Pressel, 1906)

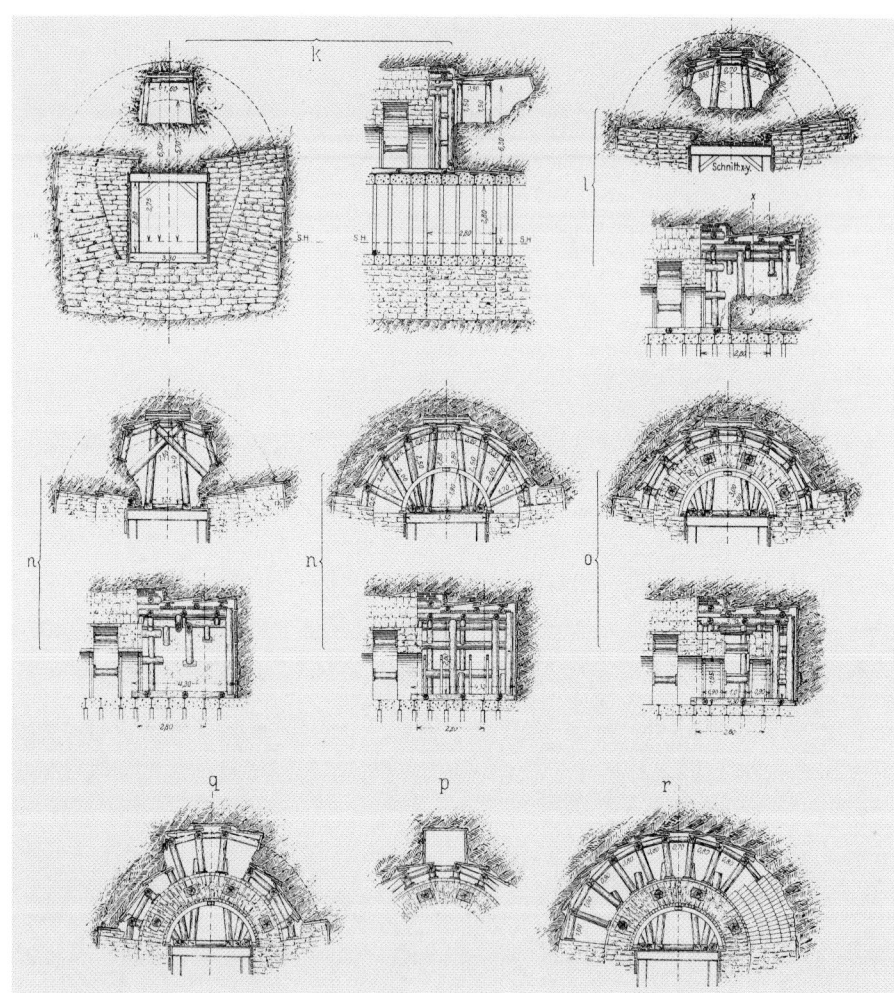

△ The excavation of the stretch of squeezing rock lasted a whole year. In the process the free unrestricted operation of the railway trucks with more than 40 trains every 24 hours had to be maintained. The pictures show the working procedure in the roof section in squeezing rock in Tunnel I on the south side.

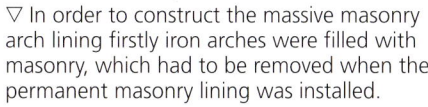

▽ In order to construct the massive masonry arch lining firstly iron arches were filled with masonry, which had to be removed when the permanent masonry lining was installed.

△ The best solution was to use sturdy iron supports, which fulfilled the requirements. These frames comprising I-beams were installed side by side, the intermediate gaps initially being filled with timber and then with concrete. Driving through this 42 m section took nearly 7 months. Even this heavy iron construction was observed to suffer considerable deformations.

Water Inflow

The big delays on the south side were also due huge water inflows.

"At a distance of 4430 m from the south portal an underground spring was encountered, which due to the force and intensity of its flow (it was estimated to be 150 litres/s) work had to be stopped immediately. All the drilling equipment had to be abandoned. Since this part of the excavation was no longer accessible efforts were made to open up the system of fissures in other places, of which it seemed only a branch had been encountered, thus allowing the water to escape freely. 30 m behind the working face a lateral drift was constructed from both tunnels and as a result an even bigger source of water with a flow rate of 150–200 litres/s was encountered. Also in Tunnel II, in which the excavation work could be continued, strong flowing springs were encountered at 4442 m, which caused great difficulties but did not bring the excavation work to a standstill; new springs were always being encountered."
(Pressel, 1906)

△ To dewater the tunnel a big channel had to be cut

△ Sealing the arch in stretches affected by large amounts of water infiltration

Legend   A 1 mm thick iron sheet
B Concrete
C Dry backfill behind lining
D Spring source

▷ Altogether in the stretch between 4.4 and 4.5 km from the south portal the total water flow reached 200 litres/s at 12 °C. One of the springs was utilised as a source of cooling water and it attained pressure exceeding 6 bars.

Rock Temperatures
Cooling Systems

"The temperature of the rock rose rapidly from about 40 °C at distance 6340 m into the tunnel at the start of the period of observation to about 52 °C at 7300 m. From there the increase in temperature was more gradual until at about 9100 m from the north portal it reached its highest value of 56 °C. ... The time had come to utilise the equipment planned to remove the heat transmitted from the rock and the ground water to the air in the tunnel, despite the high costs. ...

For this purpose two Sulzer high pressure centrifugal pumps with special 300 h.p. driving turbines were installed in the pump house. Each pump had a capacity of approximately 80 litres/s water flow at around 22 bar. A 253 mm internal diameter high pressure pipe was fitted to the pumps. The pipe was placed in Tunnel II and in the same way as both drilling water pipes they were insulated against the heat using charcoal enclosed in a metal sheath."
(Pressel, 1906)

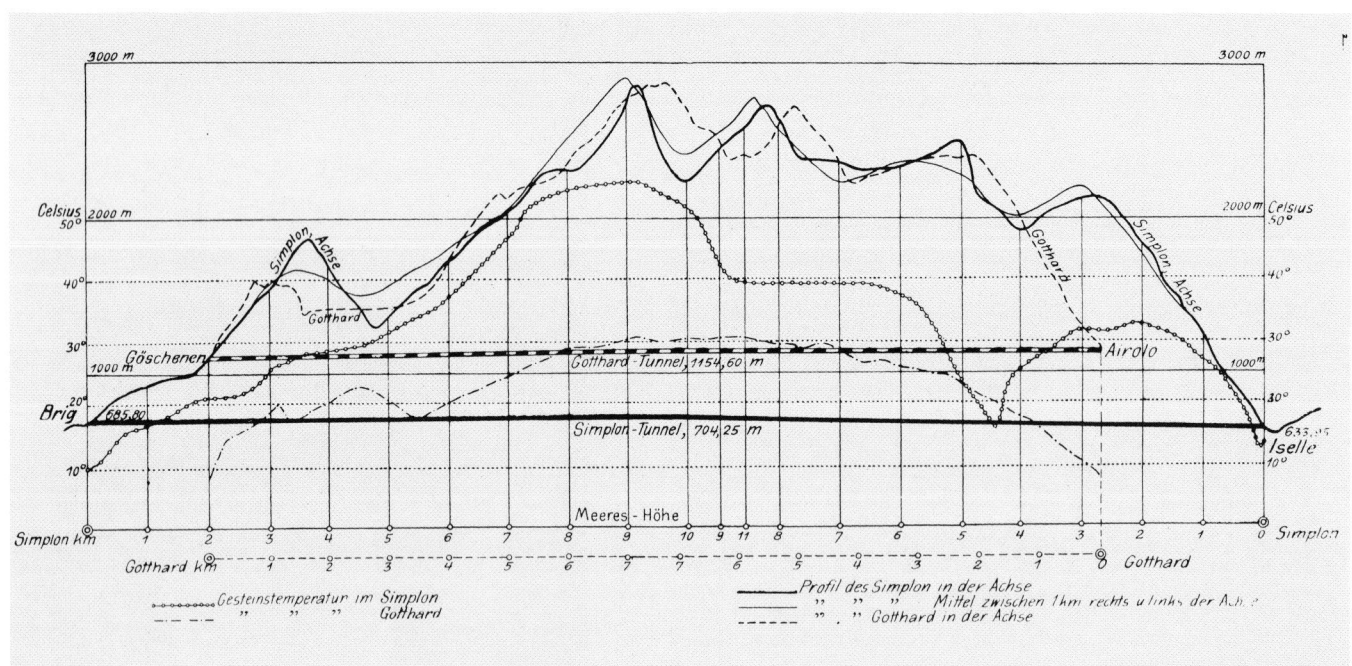

◁ Water from the cooling pipe was converted to fine spray using special valves to cool as far as possible the air in the tunnel. When necessitated by the presence of hot springs the ground water was mixed with the cooling water at the point of outflow.

▽ "The well insulated cooling tanks which were provided with axles, were placed in the flow of brine in the cold water production plant so that the tubes were flushed around and the water inside the tubes changed to ice. The mobile cooling tanks were then emptied of the brine and brought by rail into the tunnel. There they were finally connected to the pipe suppling air to the excavation."
(Pressel, 1906)

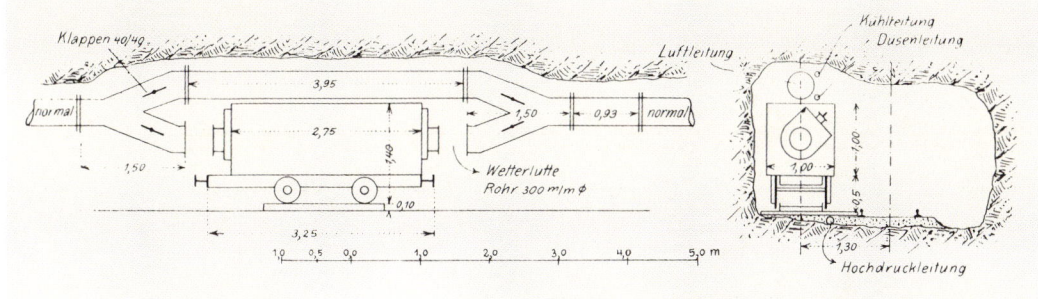

▽ On the roof: high pressure water pipe (100 mm diameter, 10 bar pressure for operating the drilling machines), cooling water pipe (250 mm diameter); on the right the charcoal insulated pipes

▷ Machine house with air compressors and water pumps

Tunnel Ventilation

"It was mainly the problem of ventilation that occasioned the firm Brandt, Brandau & Co to adopt a new approach. The employment of new construction methods and their advantages may be summarised as follows:
Using more powerful ventilators up to 50 m³ of air per second were to be blown into Tunnel II. For this purpose the mouth of the tunnel is closed with a moveable gate. Of the cross-tunnels 200 m apart connecting the main tunnels only the front two are left open. Those further away are either bricked up or closed by means of gates. In this way the air blown into Tunnel II is forced through the first two cross-tunnels into Tunnel I and through this out into the open."

(Vom Bau des Simplontunnels, 1921)

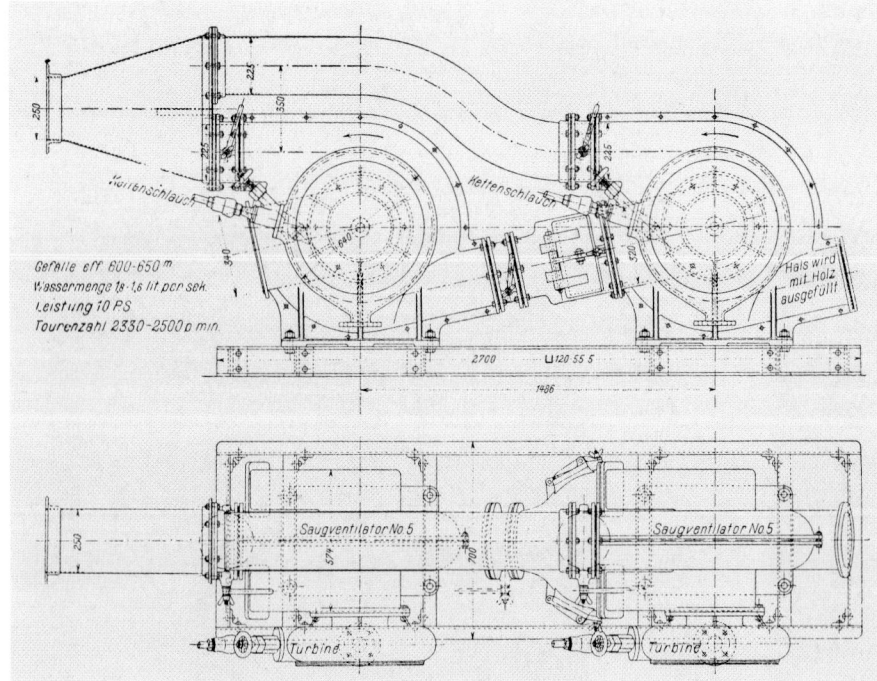

◁ "Due to the large size of the ventilator the working face cannot be supplied with fresh air so special measures were necessary. Therefore, on both sides in Tunnel II just before the last cross-tunnel either water jet blowers or small ventilators had to be installed. These separate ventilators are driven directly by Pelton type turbines, which are fed with water from the compressed air pipe."
(Pestalozzi, 1904)

▷ Tunnel I with the pipe for
the tunnel ventilation

▽ North portal of Tunnel I
with a wooden ventilation
channel, through which the
air was blown into the tunnel

Passing Place in the Middle of the Tunnel

"At 8808 m from the portal the north end of the planned passing place was reached, which was constructed to 9446 m from the portal and included Tunnel II. The cross-section of Tunnel II was increased to be the same as that of Tunnel I over a length of 527 m. Between both tunnels two connecting sections for the passing of the trains had to be built. Each of these sections consisted of $2 \times 14 = 28$ rings of gradually increasing width and height. The largest width dimension of the excavated cross-section amounted to nearly 15 m, the greatest height dimension to 8.3 m, and the lengths of these two rings were 3.25 and 3.80 m, respectively. We are making special mention of this fact because in squeezing rock conditions, in which the tunnel enlargement for the passing place was largely situated, particular care has to be taken. In the area of the enlargement with a 0.2% increase in size towards the south the tunnel base heaved so much over long stretches that water could not flow freely towards the north, but had to be pumped from the freely forming sumps. All water arising from the excavation work had to be carried away in pipes located above the working section in the enlarged stretch of tunnel for the passing place."
(Pressel, 1906)

◁ Supports for the construction of the masonry lining

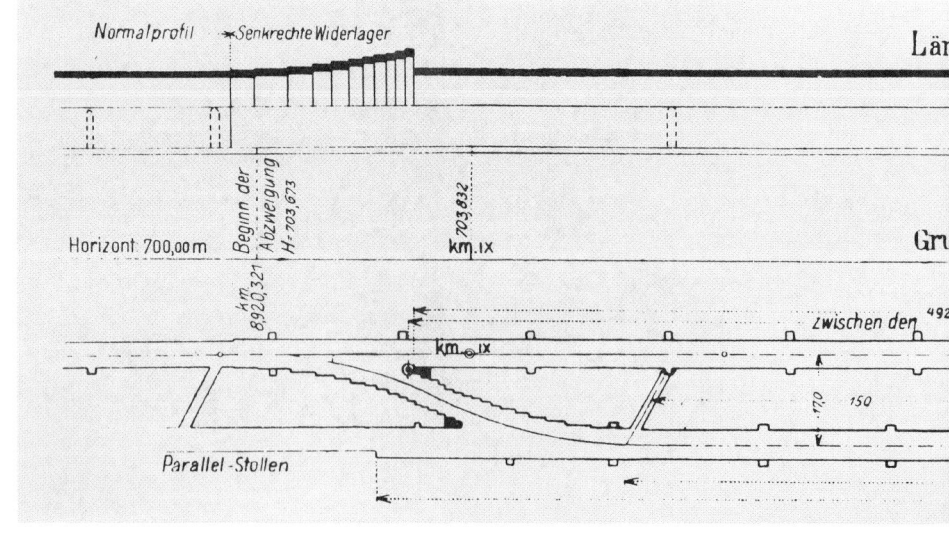

▷ Passing place and signal station inside the tunnel

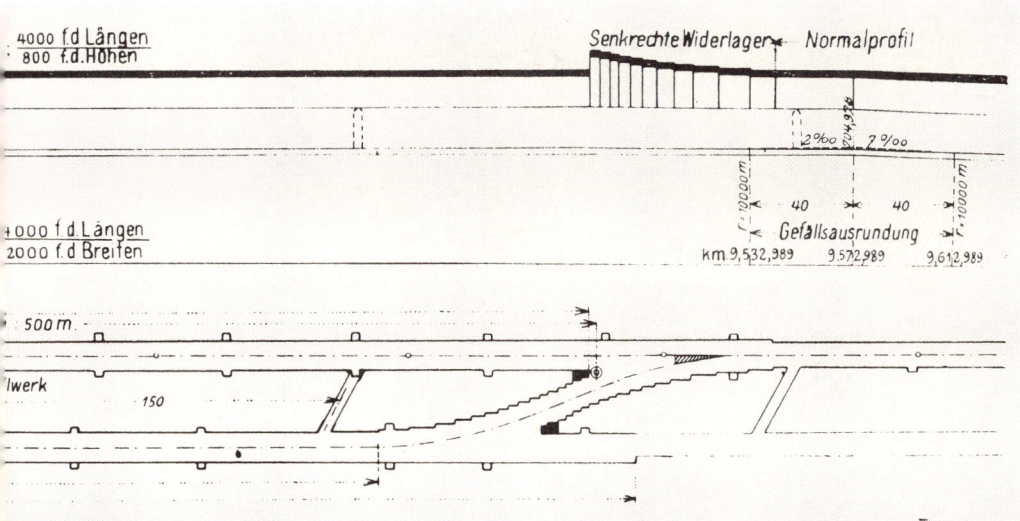

4000 f.d.Längen
800 f.d.Höhen

Senkrechte Widerlager   Normalprofil

2 %o   7 %o

r = 10000 m   r = 10000 m

40   40

Gefallsausrundung

km 9.532,989   9.572,989   9.612,989

4000 f.d.Längen
2000 f.d.Breiten

500 m

lwerk   150

◁ Passing place in the middle of the tunnel, overall plan and longitudinal section

Breakthrough

The breakthrough of the 24th February 1905:

"On the morning of the 24th it was agreed that they should travel into the tunnel at 7.25 a.m. When they heard over breakfast that in the night they had progressed 2.50 m in a single attack, whereas under normal drilling conditions usually only 1.50 m were excavated, they had to reckon with the possibility that in the next attack the breakthrough would occur. In such a moment the workers are not to be stopped; every shift wants to have the honour of holing through; they work at fever pitch with the excitement, the holes are drilled deeper than to plan and the explosive charges are doubled in some parts of the rock face.

There was an unfortunate occurrence in that shortly before they were due to depart a material train derailed about 1000 m from the portal; one had to wait until the track had been cleared. Then just before 8 a.m. there was a telephone call from the tunnel, the breakthrough had been achieved. And before they had time to enquire about the details the narrow mountain valley echoed with the hooting of the engines and from the working places transmitting the joyful news throughout the area. Within moments the windows and other places were decorated with flags."
(Rosenmund, 1905)

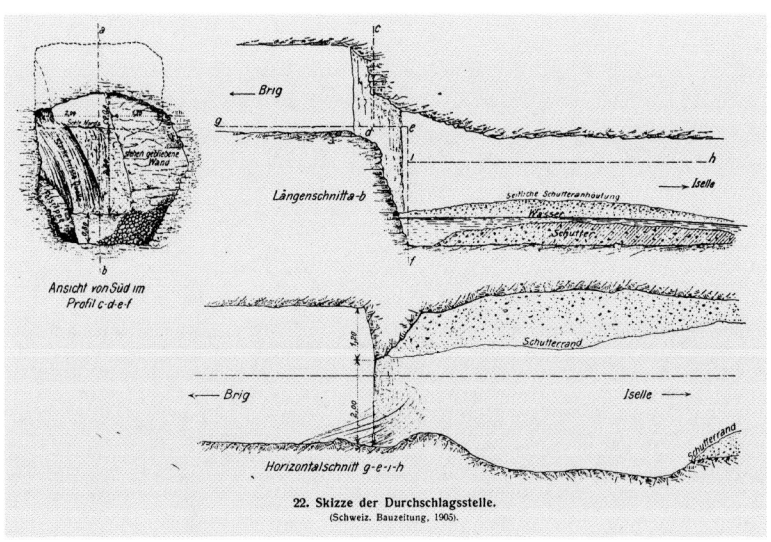

22. Skizze der Durchschlagsstelle.
(Schweiz. Bauzeitung, 1905).

△ Sketch of the breakthrough area

During drilling of the holes there was no trace of water; after blasting they pause for a moment and they were just about to go and inspect the results of the blasting when there was a loud noise and the water flooded in, nearly 80 cm high."
(Lecture of Prof. Dr. M. Rosenmund)

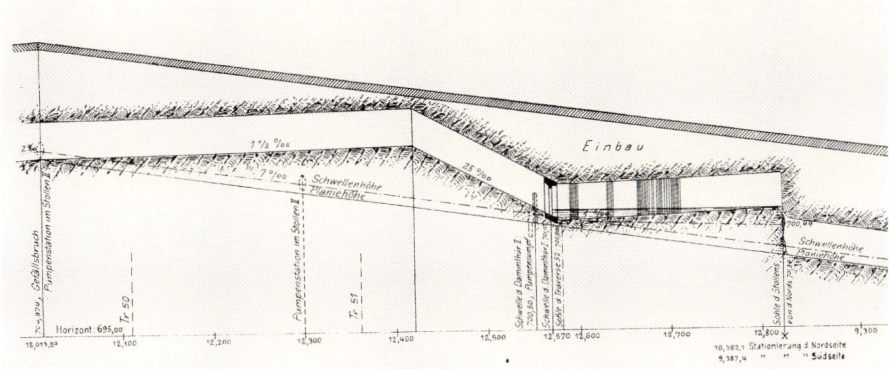

△ Due to the delays in construction on the south side they drove the pilot tunnel on the north side beyond the planned point of culmination. As long as possible it was done with a 0.15% gradient, then descending with 2.5% in order to continue with a 0.15% rise. All the water from the hot underground springs had to be pumped up over the descending stretch. The workings were drowned many times but could be dried out again with the help of massive additional equipment until on the 18th of May 1904 again a large spring was encountered and the excavation work had to be stopped. It was then decided to excavate the remaining 1 km to the point of breakthrough from the south side.

△ Postcard commemorating the breakthrough of the Simplon Tunnel.

The Lötschberg Tunnel

Total length	14'605 m
Constr. period	5½ years
Opening date	28th June 1913
Total costs	50.3 mio. SFr.

△ Breakthrough of the top heading in the Lötschberg Tunnel (from left to right Mr. Wiriot, Mr. Prud'homme, Mr. Moreau, Mr. Rothpletz, Mr. Zurcher)

Project and Geology

Project

"The Lötschberg Tunnel as the most important structure of the Bernese Alpine Railway connects Kandersteg on the north side with Goppenstein on the south side under a range of the Bernese Alps between Balmhorn and Hockenhorn. The original alignment was planned as a single straight line … but due to the major collapse under the Gastern valley a change of alignment was decided upon. Between tunnel km 1.200 and km 9.600 it was shifted eastwards from the straight line. The resulting tunnel length between the already constructed portals is now 14.605 km." (Schlussbericht, 1914)

Geology

"From the geological viewpoint the longitudinal profile can be divided into four zones in the following sequence from north to south as described below:
– …Sedimentary rocks of the Doldenhorn-Blümlisalp group, length of zone 3.970 km
– …Massif of the Gastern granite, length of zone 6.986 km
– …Zone of crystalline shale, length of zone 3.297 km."
(Schlussbericht, 1914)

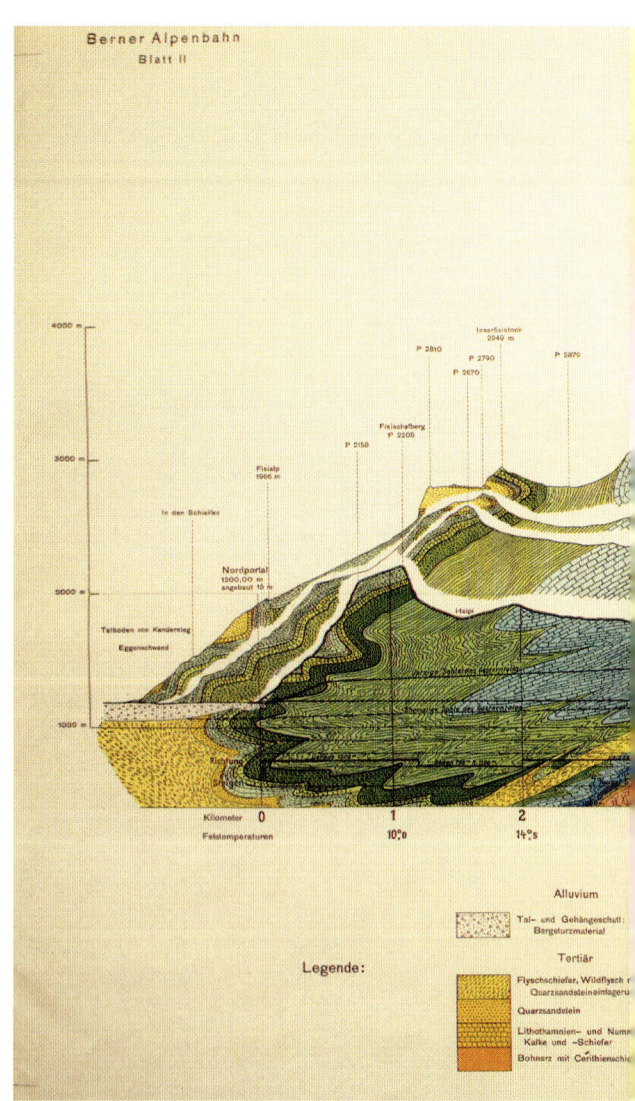

▷ Geological longitudinal profile from the final report on the construction of the Lötschberg Tunnel of the Bernese Alpine Railway

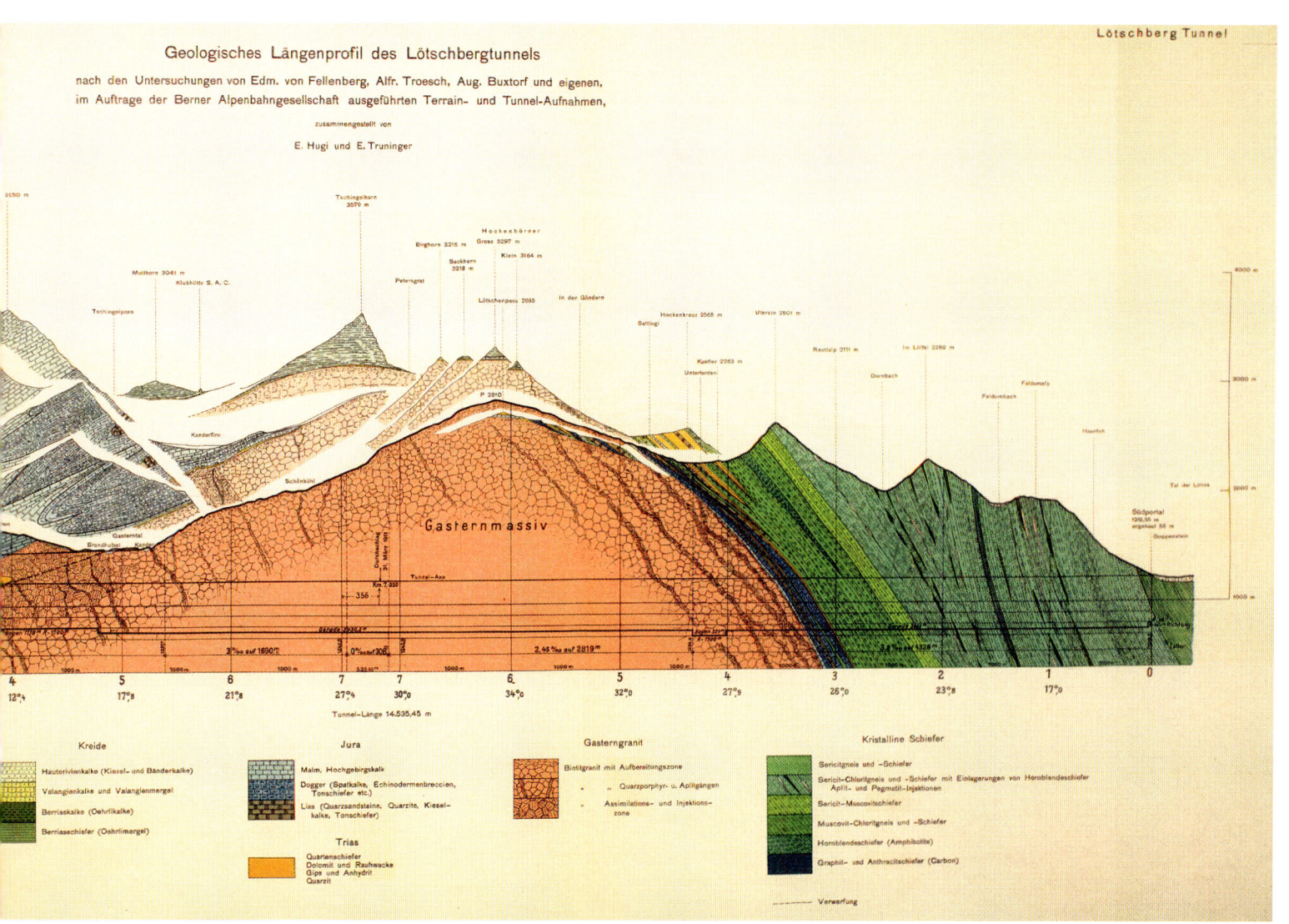

Construction and Service Railway on the North Ramp

Service Railway on the North Ramp

"The relatively high altitude of the Lötschberg Tunnel…necessitated an expensive and time-consuming transport of the equipment needed for the tunnel construction from the nearest railway stations at Frutigen and Gampel to the tunnel entrances… Since an uninterrupted operation of the service railway was of paramount importance, the contractor decided to build the service railway with its own track thus avoiding in general the alignment of the main railway line."
(Berner Alpenbahn, 1907)

Construction Railway on the North Ramp

Since for the service railway a completely independent track was laid from that of the final railway, a construction railway was necessary, which followed the alignment of the various construction sites of the final railway line. On the north ramp the construction of the required 12 tunnels was begun in the year 1910.

△ The wooden Schlossweide viaduct of the service railway was 155 m high.

"For the construction of the… (service railway) a track gauge of 0.75 m with a maximum gradient of 6‰ and a minimum radius of 50 m were specified."
(Andreae, 1940)

◁ Service railway, north ramp: Aegerten viaduct with a material supply train going to Kandersteg. This structure of 217 m length and 19 m height was the highlight of the service railway. This type of bridge construction resembled that of the American forest railroads.

◁△ Two construction sites along the north ramp, which were accessed by the construction railway.

Service Railway on the South Ramp

Providing access to the south portal of the Lötschberg Tunnel and the construction sites along the south ramp:

"The contractor chose the solution of a service railway following in its main axis the alignment of the railway line to be built from Brig to Goppenstein. Only where long tunnels... could be fairly easily avoided and in deeply cut side valleys crossed by the main railway line carried on big bridges did it have its own alignment."
(Andreae, 1940)

These extra loops resulted in large additional lengths and in part also difficult construction work.

"The service railway passed altogether through 36 tunnels of length 5310 m of which 13 alone accounted for 4070 m and formed part of the future railway line. ... The travel time from Brig to ... Goppenstein, with the servicing of the numerous stations and construction sites amounted to 3¹/₂ hours for the upward journey and 3 hours for the downward one."
(Andreae, 1940)

△ Large and difficult construction components could not be transported by the service railway. Their transport made use of horses, whereby the teams of horses sometimes had to be separated in the curves.

"The horse track to Goppenstein to the south entrance to the great tunnel was widened and corrected; however, it could only be used occasionally because as mentioned above it was closed for a long time due to avalanches." (Zollinger, 1910)

"At the beginning the workers had to be let down using ropes and sometimes they were suspended in the air as they drilled the first blasting holes, until the blasting work gradually permitted conditions of being able to stand."
(Andreae, 1940)

△ Stretch of the service railway in steep rocky ground in the Bietsch valley

▷ Service railway viaduct of the Mundbach stream

Slides during the Construction of the Service Railways

Slides in Soil and Rock

North ramp:

On the north ramp, above all earth slides and rock falls in the region of the Ronenwald valley and at Fürthen caused big problems.

South ramp:

"It is clear that the unfavourable location of the railway in relation to the strike and dip of the rock layers presented in some places a potential danger; when cut through the rock slabs sloping to the horizontal and lying almost parallel to the axis of the railway line and to the bedrock surface, slides inevitably followed."
(Andreae, 1916)

"For safety reasons, in many places the slopes of the cut in the mountainside were designed to follow the bedding of the rock, which led to a large increase in the volume of rock to be removed. Where such a safety measure for the rock slope was too costly, high and thick revetment walls were necessary."
(Andreae, 1940)

◁▽ Service railway, north ramp: Slope stability problems in the area of the rock fall gallery in the Ronenwald valley. On the 17th May 1911 work on the service railway was interrupted and its track severely damaged.

▷ Service railway, south ramp: Rock slide at Hohtenn, which occurred on the 7th January 1913.

◁ Service railway, south ramp:

"The most difficult tunnel of the south ramp was the Victoria tunnel in the Baldschieder valley. This short 28 m long tunnel passes through the Victoria block, so named by the engineers because its shape reminded one of the head of Queen Victoria on the old English postage stamps. This block was part of an old rock slide. ... An exploratory shaft at the foot of the block as well as the experiences gained from the service railway tunnel led one to expect that the block was sufficiently strong and stable to carry the railway. These expectations were fulfilled, but the work proved to be very difficult."
(Andreae, 1916)

Infrastructure in Kandersteg

"On the north side the installations comprise 31 buildings with a total covered area of 8350 m². To house the administrative staff and workers 10 residential buildings and rows of barracks providing accommodation were built with an area of 2882 m²... In the installations area 8510 m of rail track of 75 cm gauge were laid with 70 points and a turntable of 5 m diameter. The rolling stock consisted of 5 steam engines, 5 compressed air engines, 380 box-type trucks, 20 platform trucks and 16 passenger trucks for personnel. The machines in the north side installation area are electrically driven."
(Zollinger, 1910)

△ The installation area at the north portal of the Lötschberg Tunnel with workshops, machine houses, electrical power house (transformers, etc.), compressor plant, ventilation plant with centrifugal ventilators, drill smithy, sawmill, stone crusher, cement works, dynamite storehouse, timber and coal bunkers, warehouses, offices, accommodation buildings, baths, fire service building, hospital, etc.

▽ "Missione Cattolica Italiana" in Kandersteg

▽ Field hospital in Kandersteg

△▷ Workshops in Kandersteg

Infrastructure in Goppenstein

Technical installations

- machine houses
- equipment for mechanical drilling
- compressors to produce compressed air for the mechanical drilling and the compressed air engines
- transformers
- air tanks
- on the Goppenstein side a total of 17 km of track was used for transportation inside and outside the tunnel.

Workers and social structures

"The workers were mainly Italians, 40% of them from southern Italy, 30% from central Italy, 12% from Lombardy, and 15% from Piemont, while the remainder were Swiss. A large number of the workers had their families at the site so that in Goppenstein the guest workers' population grew to 3600 persons over the construction period." (Schlussbericht, 1914)

Amongst others the social structures for the workers were:

Accommodation for the supervisory staff and workers; buildings with canteens, bath and shower rooms (showers for the workers, baths for the supervisory staff), drying rooms for the tunnelling clothes, washing room, disinfecting system, hospital for 25–40 sick or injured persons, 2 bakeries, 2 butcher's shops, 5 grocery shops, school, post office, police station, hotel run by the company, 1 cinema.

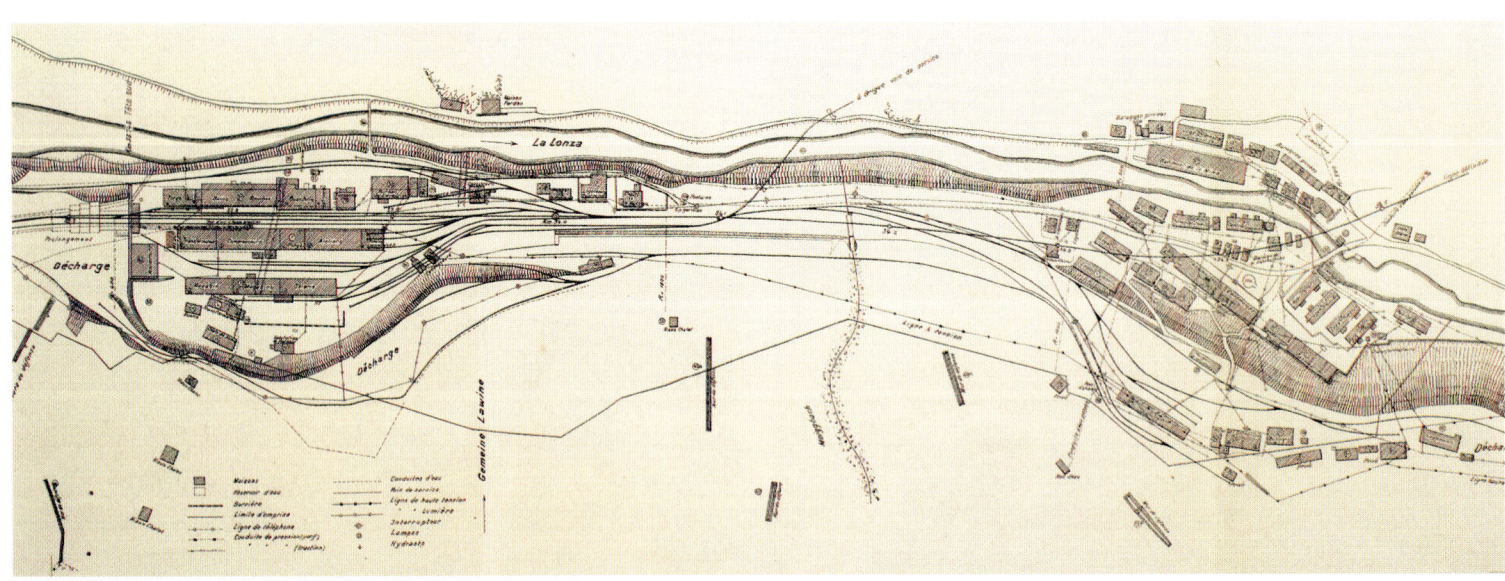

△ Service railway bridge in Goppenstein in spring 1909. Train with two steam engines, a passenger truck, eight tip-up trucks and a flat truck

◁ Plan of the installation area, with the service and construction railways, infrastructure facilities as well as waste dumps (Schlussbericht, 1914)

▷ Accommodation buildings for the tunnel construction workers; behind are the dumps for the tunnel excavation debris.

Surface and Underground Surveying

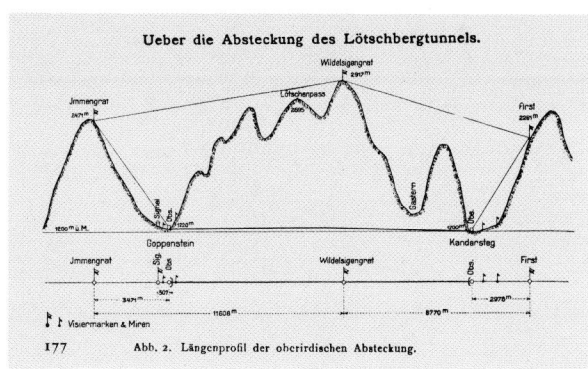

Ueber die Absteckung des Lötschbergtunnels.

177   Abb. 2.   Längenprofil der oberirdischen Absteckung.

Triangulation

"In the summer of 1906 the Lötschberg Railway Construction Company commissioned the surveyor T. Mathys with the work of surveying the Lötschberg Tunnel. Due to the short time available Mathys restricted himself provisionally to a simple triangulation, which connected the two points of the tunnel specified by the company. Departing from the method used in the Gotthard and Simplon tunnels the two end points of the tunnel were not connected by an independently carried out triangulation, but instead on the basis of the triangulation of III degree of the Bernese Oberland produced by the Swiss Ordnance Survey."

"From the coordinates of the points the azimuths of the tunnel direction and of the connecting directions of the axis points Kandersteg and Goppenstein were calculated, from which the angles could be derived, which had to be set out from these directions in order to control the tunnel direction in relation to these two axis points."

"In Kandersteg a very solid observation pillar was constructed of concrete... and subsequently called 'Trig Point' Kandersteg. To protect the instruments and the surveyor it was covered by a simple wooden hut. ... The 'Trig Point' Goppenstein was also covered by a wooden hut."
(Baeschlin, 1911)

Surveying on the ground surface

"On making a reconnaissance of the triangulation network Mathys realised that a direct surface survey of the tunnel axis was possible. He proposed that on the north and south sides of the tunnel one point on the tunnel axis could be chosen in such a

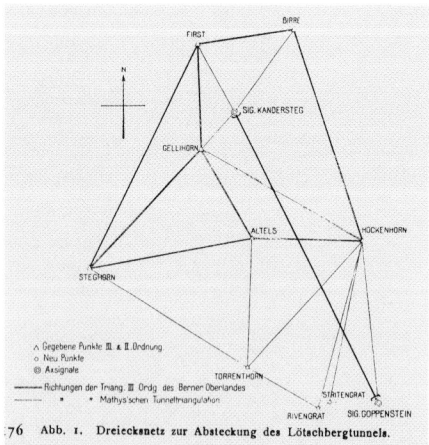

76   Abb. 1.   Dreiecksnetz zur Absteckung des Lötschbergtunnels.

way that from it both the corresponding triangulation point and the axis signal as well as the axis point on top of the mountain ridge could be seen."

"From the axis signal Kandersteg a measuring staff at the position of the provisionally fixed axis point on the tunnel roof could be read. Then from the axis signal Goppenstein by measuring angles... the triangulation point on the Immengrat could be brought onto the axis. There the theodolite was placed exactly on the axis and the measuring staff, which is graduated on both sides, placed on the Wildelsigen ridge could be observed. According to Mathys the two positions of these triangulation points differed by 2.5 cm."
(Baeschlin, 1911)

Underground survey control

"For a straight tunnel axis, after excavating about 600 m the position of the points fixed so far were controlled by means of a primary survey control. During these controls the work in the whole of the tunnel was stopped, so that no disturbance was caused to the

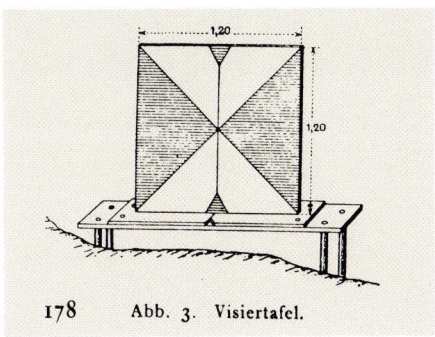

178   Abb. 3.   Visiertafel.

measurements by material transport or smoke. Thus usually a public holiday was chosen for this work. ... Normally during such a primary survey control the following operations were performed: checking direction, measuring levels and lengths. ... The instruments used were basically the same as those used for the Simplon Tunnel."
(Baeschlin, 1911)

The first survey of a large curved alpine tunnel

The catastrophe of the 24th July 1908 below the Gasteren valley led to the company's deciding upon a "new alignment to avoid the difficult places, by introducing three curves each of 1100 m radius". The total length of the curves was 2267 m. Since the previously constructed major alpine tunnels were straight this was the "first case of surveying and setting out a large curved alpine tunnel".
(Baeschlin, 1911)

As a result Prof. F. Baeschlin (of the Swiss Federal Institute of Technology, Zurich) was asked to provide expertise, on "whether from the standpoint of surveying it was possible in view of the curves to maintain the specified direction with sufficient accuracy, such that a lateral breakthrough error of less than 1 m could be guaranteed".

Prof. Baeschlin responded to this question in the affirmative:
"To set out the curves a polygon with about 100 m sides was defined by hectometre pegs, measured as accurately as possible, in that both the included angle and the lengths of the polygon sides were measured. As change-over points for the theodolites and the sighting lamps iron tripods were used, which were left in position until the measurements had proceeded some distance".
(Baeschlin, 1911)

It should be observed that "due to the three curves defining the tunnel special care had to be taken with longitudinal measurement".
(Schweiz. Bauzeitung, 1911)

"For the longitudinal measurement a pair of narrow wooden measuring boards of 5 m length were used. ... Each board was provided with a measuring device at the end. ... The length of the boards was determined exactly on a reference comparator before and after each measurement."
(Baeschlin, 1911)

"Whereas for the curve on the south side the tangent point was made accessible by two short tunnels and the control survey was carried out directly

from here with the aid of an angle, on the north side this was impossible. Here one was forced to set out the construction axis as a circular arc..."
(Schweiz. Bauzeitung, 1911)

Accuracy of breakthrough

"The checks after the breakthrough showed a deviation of:
– 25.7 cm in the tunnel direction
– –41.0 cm in the length
– 10.2 cm in the height."
(Schlussbericht, 1914)

"We are extremely pleased about the excellent results of this part of work and wish to express our congratulations to Prof. Baeschlin; we are sure that in this we speak for all our colleagues as well as the Federal Institute of Technology, Zurich!"
(Schweiz. Bauzeitung, 1911)

△ This picture shows Prof. Rosenmund during the surveying of the axis in the Simplon Tunnel.

Sequence of Excavation in Cross-Section

Position of the pilot tunnel

"In the case of the Lötschberg Tunnel, as also of the Simplon and Mont-Cenis-Tunnels, but in contrast to the Gotthard Tunnel, the pilot tunnel was driven as a bottom heading, i.e. such that the lower edge coincided with that of the final tunnel and likewise with the axes."
(Andreae, 1940)

Driving of pilot tunnel

"The blasting operation was carried out as follows: firstly holes were drilled by machine into the working face of the pilot tunnel, then filled with powder explosives and detonated, whereby the drill carriage with the drilling machines had to be moved back some distance. After blasting the debris was removed beginning with the track area, so that the drilling work could be continued as soon as possible.

Up to this point mucking took on average 22 min/m³, … During drilling the remaining spoil material was removed. While the spoil material was being removed the machine operators carried out the necessary additional work, such as clearing the roof and side walls, laying the track and the service lines, etc.

The drilling, blasting, mucking, progressed continuously during the whole of the working time, without a break at change of shift. Each new crew carried on where the other one had left off."
(Andreae, 1940)

▷ Passing place in the pilot tunnel of the Lötschberg Tunnel

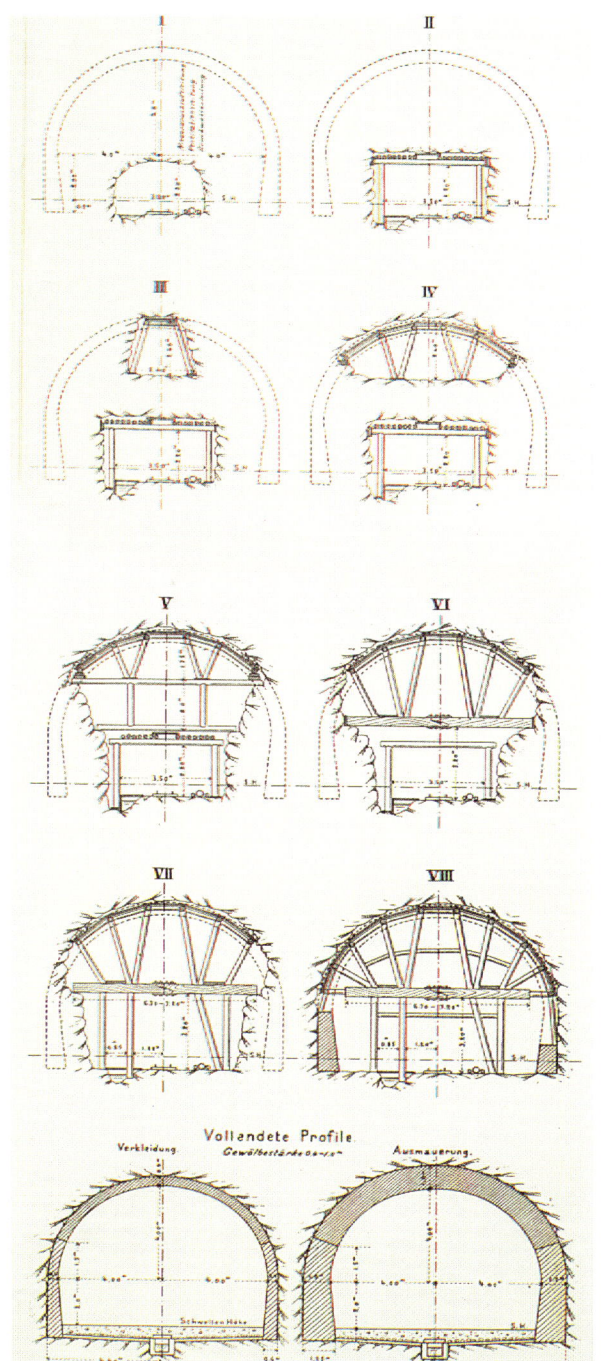

△ Excavation crew with simple tools like drills, picks and shovels

◁ The method of excavation with the pilot tunnel in the bottom
(Schlussbericht, 1914)

Full Face Excavation and Masonry Lining

Full Face Excavation

"The full face excavation was begun with the top heading and on the north side this was only carried out manually to a distance of 459 m; by far the greater part of 7076 m was excavated from the bottom heading by means of slits machine drilled from the bottom heading. On the south side of the 7000 m of top heading 515 m was excavated by hand, 5295 m with mechanised drilling in various attacks and 1190 m by means of slits drilled from the bottom heading.

If the top heading method is employed, firstly the roof is excavated followed by both side headings. ... During excavation to the full profile 30–40 drills and drilling hammers were in operation. Great savings were made through the shift working."
(Schlussbericht, 1914)

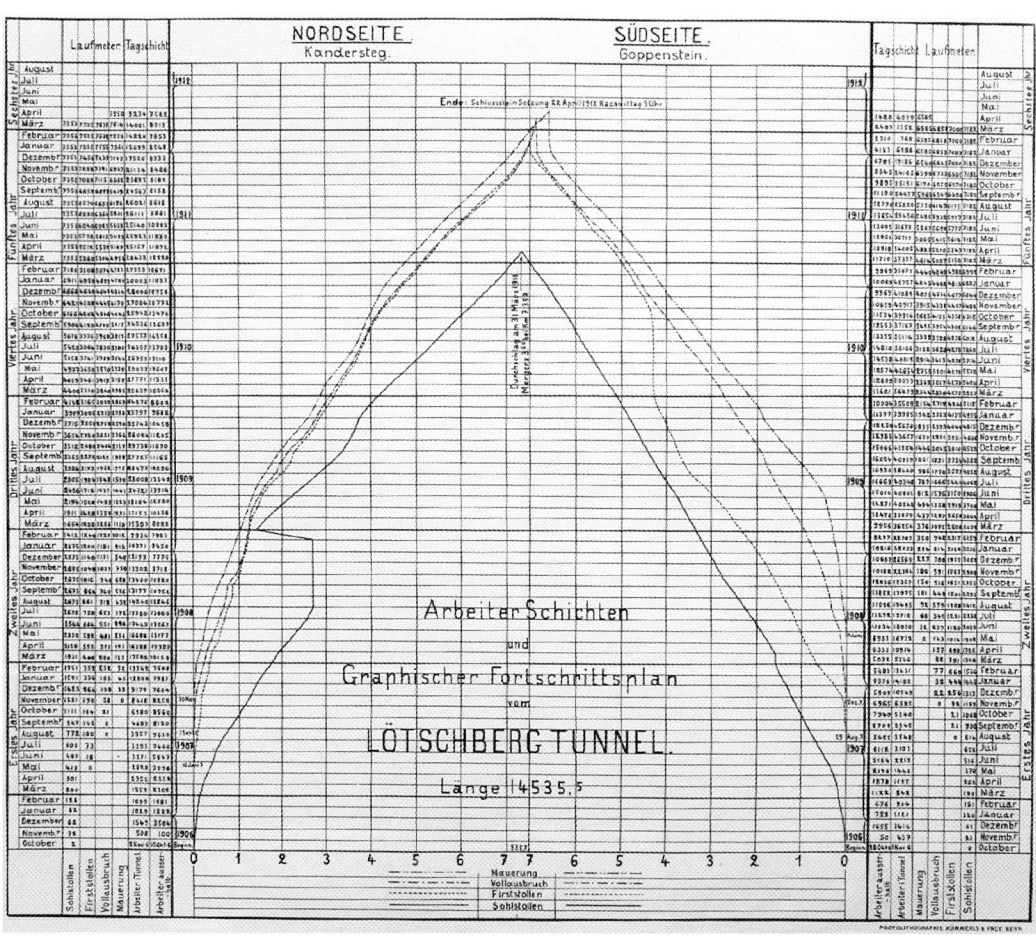

△ Work schedule for the Lötschberg Tunnel: plan for the period between start of construction in October 1906 to the completion of the structural work in April 1912 (Schlussbericht, 1914)

84

▷ Mucking using tipping trucks. The material was loaded into the trucks using short shovels. In the background the roof support installation may be seen, with a total of six roof beams being visible.

Mucking

The spoil material was removed manually. For the bottom heading, of the 17–20 workers per shift 8–10 were used for mucking. "All attempts to use mechanical spoil removal proved to be too time-consuming; one was forced to use manual labour, which could be adapted better to the conditions – one only needed a good organisation of the workers and close supervision."
(Schlussbericht, 1914)

The masonry lining

"Following the excavation of the full profile came the construction of the masonry lining, which began with the haunches and finished with the roof arch. In the stretches with an invert this was constructed after the roof had been completed. The lining was carried out in rings of length varying between 4 and 10 cm, depending on the rock pressure conditions. The invert was constructed over a length of 2691 m."
(Schlussbericht, 1914)

▷ Masonry lining after excavation to full profile

Drilling Machines for the Blasting Operations

"For the final excavation work on the north side percussion drilling machines System Meyer (Müllheim, Ruhr) were chosen, which came into operation on the 18th June. At first three of these were in operation using a horizontal jack leg. For the permanent operation of the final compressor plant (also supplied by the firm Meyer) drill carriages with horizontal jack legs were employed, which were pressed against the tunnel sides and to which up to 5 drilling machines could be fixed. Each machine was operated by two men, one behind at the controls for forwards and backwards movements, and the other at the front for positioning the drill holes and changing the drill bits. At the front it sufficed to have 3 men to operate the 3–5 machines. When working with four machines the crew consisted of 1 supervisor, 7–8 machine operators, 8–10 men for mucking and 1 apprentice, in total 17–20 men." (Andreae, 1940)

▷ Drilling carriage with two horizontal jack legs and a total of 5 drilling machines

▷ Side view of a moveable drilling machine jack legs (above) and view from working face (Schlussbericht, 1914)

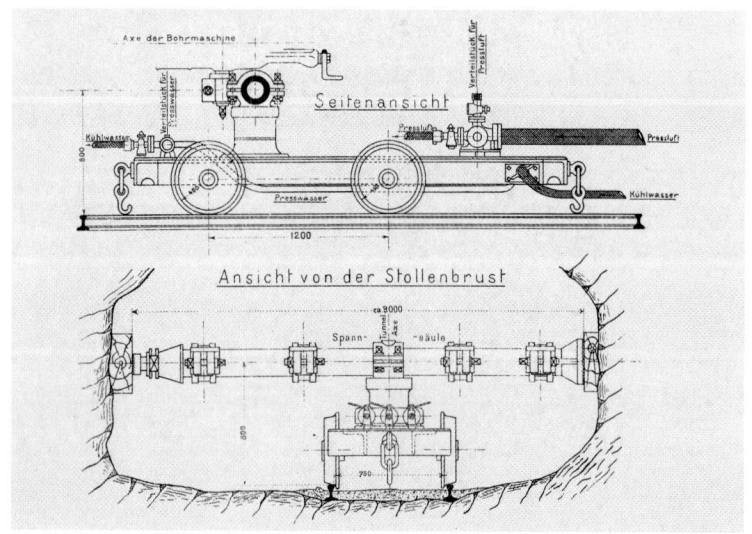

▽ Jack leg with two drilling machines

▽ Compressed air drilling machine (System Meyer) with horizontal jack leg

▷ Percussion drilling machine
(System Meyer) attached to
fixed vertical jack leg

Transport of Workers and Materials

Transportation

"In the tunnel on the north side horse transport was carried out up till the 12th September 1908, when the compressed air engines took over the transportation. On the south side the compressed air engines were put into operation on the 15th August 1908, from the portal up to the tunnel passing place; there to the working face horse transport took over. On the north side horse transport was used occasionally from the passing place to the working face. On the south side in the tunnel in addition to compressed air engines steam engines were often used for transport up to the passing place, which hardly improved the air quality in the tunnel. For transport into and out of the tunnel the workers travelled in special trains for the shifts up to the tunnel passing place. Per shift there were four trains, i.e. daily twelve."
(Schlussbericht, 1914)

The compressed air engines

"There were three different sizes of compressed air engines, i.e. with 2 coupled axles, 7 tons adhesion, an air capacity of 2.63 m³, and 40 horse power; with 3 coupled axles, 11 tons adhesion, 4.4 m³ air capacity, and 70 horse power; with 4 coupled axles, 24

△ On the south side two steam engines were used for transport in the tunnel. The picture shows such an engine at the tunnel portal in Goppenstein.

▷ Compressed air engine

tons adhesion, 8 m³ air capacity and 180 horse power."
(Schlussbericht, 1914)

The air compressors

"The air compressors providing air to the tunnel engines had to compress the air to 100–120 bars. On both the north and south sides there were two such compressors in operation. On the north side they were five-stage compressors of the firm Meyer in Müllheim (Ruhr), whereby the air was compressed successively to 2, 6, 19, 58, and 120 bar. The compressor performed 135 revs/min and compressed 16 m³ of air per minute."

▷ "To load up the compressed air engines two five-stage high pressure compressors of type System Meyer, which compressed 16 m³ of air per min. to 120 bars overpressure, each with a motor for three phase current of 500 volts and 275 amps."
(Andreae, 1940)

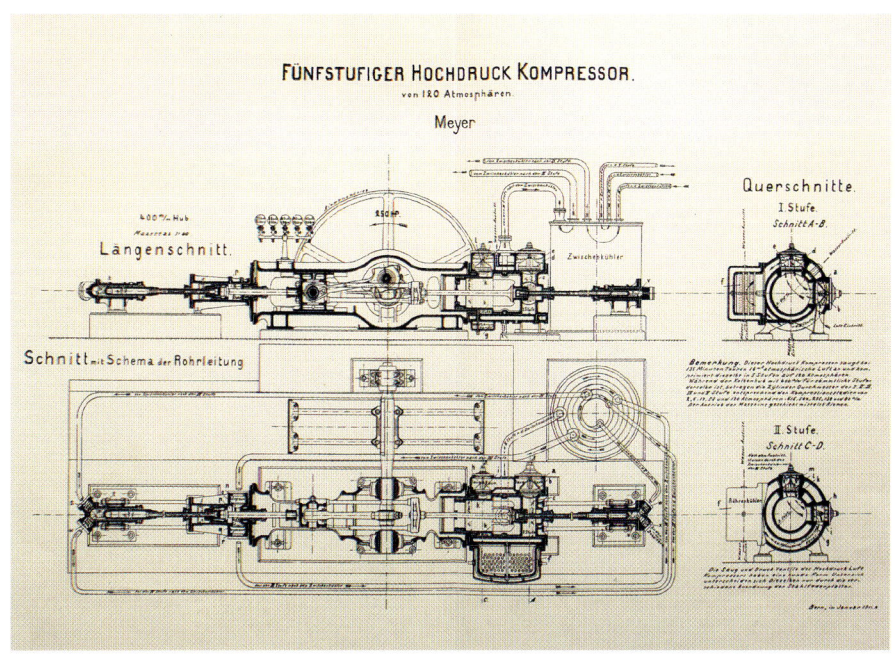

Collapse under the Gastern Valley

From the memoirs of Dr. h.c. F. Rothpletz "Recollection of the most tragic day of my life and its consequences" from the year 1944:

"On the 23rd July 1908… the working face of the bottom heading pilot tunnel reached a distance of 2.65 km from the start of the tunnel… The rock, which was permeated with calcite veins, was a very compact black alpine limestone… The bottom of the Gastern valley was situated 172 m above the pilot tunnel and through it the river Kander flows in a wild manner."

"I was at home about 200 m from the tunnel portal and at about 1 a.m. I quickly fell into a sound sleep, tired as I was, out of which I was awoken by an engineer at around 3 a.m. He said: "Something has happened in the tunnel and then he hurried off."

"When I came to the tunnel portal all was dead still. In the control office I was told that everyone had gone into the tunnel, but one did not know what had occurred. The night patrol showed that the control tabs of practically the whole crew had not been given in, so that the workers must still be in the tunnel. With a supervisor I went immediately into the deserted tunnel to find out for myself what had happened. 1200 m from the tunnel portal we came across a pile of rubble. It was soon apparent that at a distance of about 1500 m from the portal the tunnel was filled with debris, out of which some water flowed."

"It was immediately clear: collapse and inflow of loose material, probably from the Gastern valley. The whole crew lost! Investigations in the Gastern valley revealed a funnel-

Fig. 1. Coupe à travers la vallée de Gastern
Le long du tracé actuel.
1:5000

◁ Longitudinal profile from the "Rapport", 1909

24th July 1908: Excavation of the bottom heading pilot tunnel. From left to right, about 2675 m from the portal at Kandersteg. In the night of the 24th July a collapse of the tunnel face occurred with inflow of loose material and water from the Gasteren valley sediments. The tunnel was filled with material over a length of 1500 m. There were 25 deaths.

shaped collapse of about 80 m diameter and 3 m depth – that is: breakthrough of the rock face due to the blasting of the last attack... and inrush of material from the Gastern sediments into the tunnel... The assumption of the geological report, that there was still 100 m of firm limestone rock above the tunnel must be wrong."

"Not far from the end of the pile of debris the search party found one person, who was dead. As representative of his 24 comrades he was the only one to be buried in the graveyard at Kandersteg."
(Rothpletz, 1944)

△ Measurement of the funnel-shaped collapsed area at the ground surface in the Gastern valley.

▽ Funnel shape at the surface in the Gastern valley as a result of the collapse of 24th July, 1908. First measurements a few hours after the collapse, together with measurements made later. The position of the pilot tunnel can also be seen, as well as of the river Kander, and the arrangement of the trial boreholes subsequently made. (Rapport, 1909)

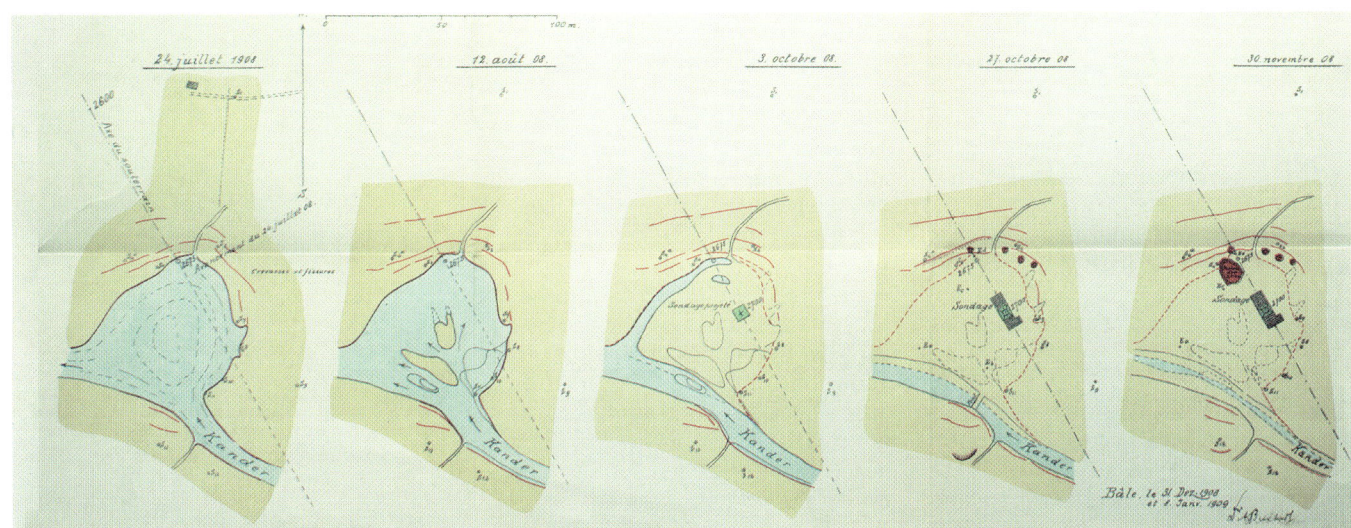

Collapse under the Gastern Valley: Assessment and Countermeasures

Exploratory boreholes and shafts

"Exploratory boreholes were required in the Gastern valley, in order to ascertain the depth of the bedrock and the extent of the loose material that would have to be tunnelled through… Finally, it was found that the Gastern valley had been eroded to a much greater depth than had been assumed and filled up with sediment material, even much deeper than the proposed tunnel and that there was not, as thought, a 100 m deep rock formation above the tunnel, but instead 172 m of saturated material from the river Kander, moraine and scree material."
(Rothpletz, 1944)

"It was decided by the company to sink two shafts together with boreholes to determine if the alternative route was situated entirely in rock and pits were dug in the bed of the river Kander."
(Schlussbericht, 1914)

Assessment and countermeasures

"Detailed studies were also necessary to determine with what measures it might be possible to tunnel through the roughly 350 m stretch of loose material in the Gastern valley and how much time needed and what costs would be incurred, and if this task could be completed at all."
(Rothpletz, 1944)

Compressed air technique

"The driving of the tunnel using the compressed air technique did not seem to be feasible because one had to reckon with a water pressure of perhaps 10 bars or more. For a pressure exceeding 3 bars, however, the limits of admissibility for acceptable human working conditions is already reached."
(Rothpletz, 1944)

Ground freezing and cement grouting techniques

"The success with cement grouting as also with the ground freezing technique was questionable, since with infiltrating water the cement like the expensive artificial freezing is lost; … A problem would also be presented by the places which with even the use of both techniques might not be sufficiently stabilised and thus there would be a renewed risk of the collapse of the roof and more fatal accidents with an overburden of 172 m of loose fill material."
(Rothpletz, 1944)

△ Material transport for the boreholes in the Gastern valley

▷ Site of borehole with covered boring tower in the Gastern valley

▽ Results of both trial borings presented in a cross-section through the Gastern valley

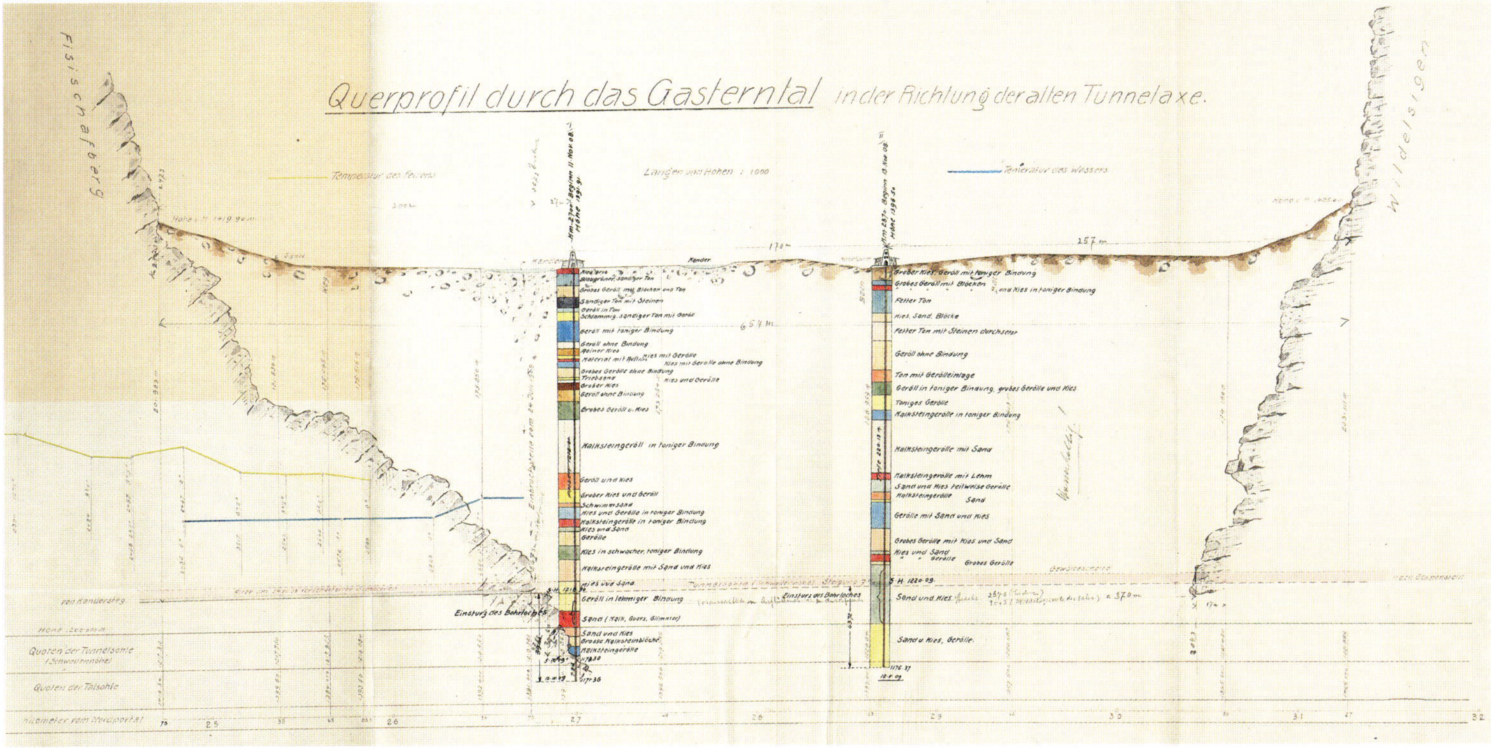

Querprofil durch das Gasterntal in der Richtung der alten Tunnelaxe.

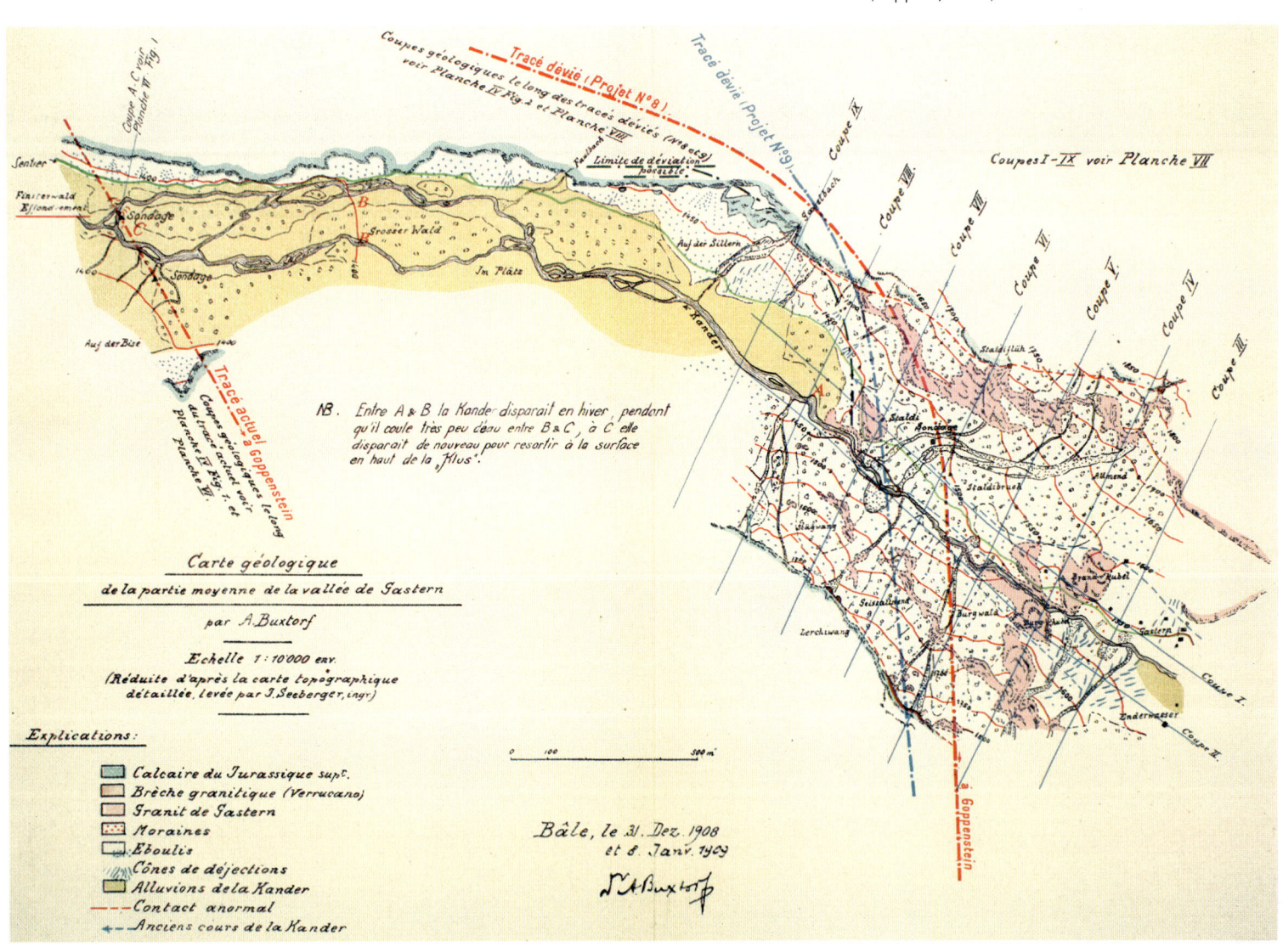

▽ The old route with the position of the collapse and the new route (Rapport, 1909)

94

Avoiding the dangerous zones

"With the immediate arrival of the seven contractors from Paris the question was raised already on the second day of a new route avoiding the accident area in the Gastern valley. ... For the safety of the workers I was against carrying out such dangerous work until some other safer method could be found to complete the excavation of the Lötschberg Tunnel, and it seemed to us, it would be better to avoid the danger zone altogether. This led to the decision to by-pass this zone accepting an 800 m lengthening of the tunnel. And this by-pass was indeed successful, with no special circumstances or difficulties arising. The tunnel, despite the considerable delays caused by this event (about 7 months), despite the extra 800 m of tunnel length and despite the 1500 m of pilot tunnel that had to be abandoned, was completed by the deadline originally set. ..."
(Rothpletz, 1944)

▷△ Ground freezing technique: "In order to obtain a solid ground above the tunnel up to the ground surface, it would have been necessary to freeze a prism ... The volume to be frozen would attain at least 1'163'750 m³."
(Rapport, 1909)

Financial consequences of the catastrophe

"The catastrophe led to a legal process over the question, who should be blamed for the financial losses incurred, the railway company or the contractor? In the main the contractor was found to be responsible. It was not an "act of God", and therefore according to the terms of the contract the contractor had to accept liability, to which he also acquiesced."
(Rothpletz, 1944)

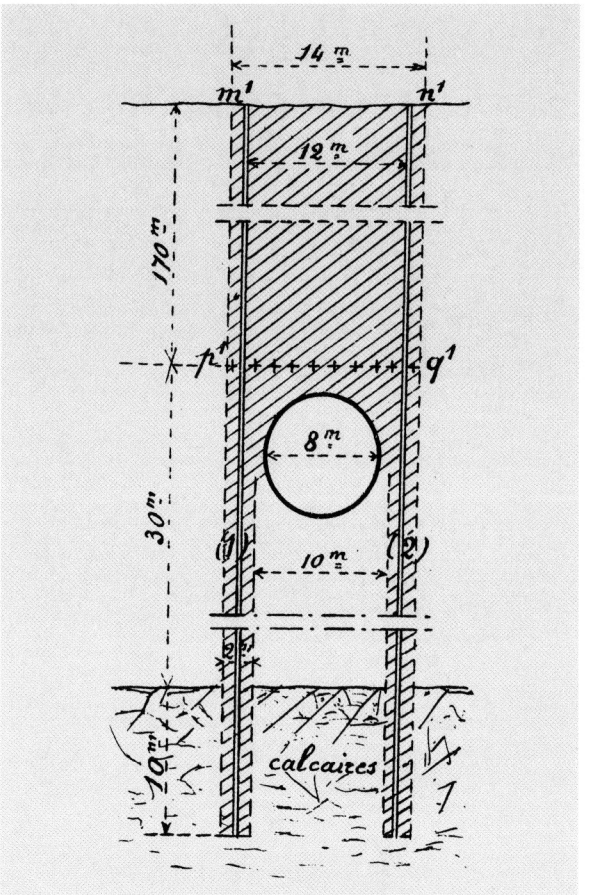

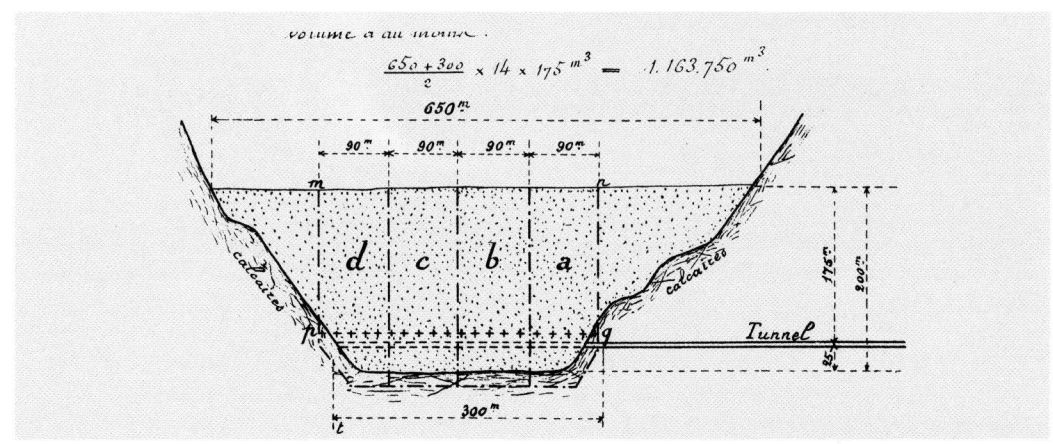

Goppenstein: the Fight against the Avalanches

The avalanche disaster in Goppenstein

"Because there were relatively few avalanches in the winter of 1906/07, despite the fact that there was so much snowfall, the engineers on the contractor's side, who were mainly foreigners and thus less knowledgeable about conditions in the mountains, were lulled into a sense of false security and despite warnings from the local authorities and local people they did not pay enough attention to the danger that avalanches presented for their installations. Some of their buildings were too lightly constructed and with insufficient resistance."
(Andreae, 1940)

"On the 29th February 1908, it snowed heavily the whole day and evening, even though there was a south west wind. The people of Lötschental, who knew their valley and its avalanches, recognised straightaway that this meant a danger of avalanches. In Goppenstein between 7 and 8 p.m. about 30 persons were having their evening meal in the hotel. Suddenly the wooden building collapsed and 11 of them were killed instantly. ... A twelfth person died some days later in hospital in Brig. ... Due to the action of the what the locals called the 'Gmeinlaui', i.e. the nasty avalanche, a powder avalanche came down,

△ Installation area Goppenstein after the avalanche of 29th February 1908. The picture shows in the foreground the snow-covered tunnel portal of the Lötschberg Tunnel; on the left one can see destroyed buildings

which hardly touched the building, ... but the wave of air pressure simply blew away the light building. With the exception of Engineer Sylva... all those persons in the hotel died, who, as was later found out by the seating arrangement, faced the oncoming avalanche, whereas those with their backs to it survived or were at most injured by the collapse of the structure. The cause of death was asphyxiation."
(Andreae, 1940)

△ The service railway buried by the "Rotlawine" avalanche in 1908

△ The mass of snow left by the "Rücklawine" avalanche in spring 1907 near the tunnel portal. The construction of the Lötschberg Tunnel was continued further through this tunnel in the snow.

△ Digging out the construction railway in the winter of 1907/1908 in the area of the "Rücklawine". Often one built a tunnel in the snow for this purpose.

In the lower Lötschental valley the danger of avalanches caused headaches in trying to fix the position of the tunnel's southern exit, the installation area and the access to the tunnelling site. It was feared there might be several avalanches, especially the "Rücklawine", the "Gmein- laui" and the "Rotlawine" avalanches. ... (Translator's note: these are not types of avalanche, but places where there was often an avalanche)

Rock Temperatures, Construction and Operating Ventilation

Construction ventilation: primary ventilation

"Two circulation systems were installed in the tunnel for the ventilation. The first, the primary one, stretched over the part of the tunnel where the masonry lining had been completed. From the portal up to the end of the finished lining a vertical separation wall was built, stretching from the base of the tunnel to the roof, and in it a fresh air duct with a clear 6.3 m² cross-section was isolated. This wall consisted of vertical steel I-section with plastered brick fill material fixed in the base and the roof." (Andreae, 1940)

Construction ventilation: secondary ventilation

"The working places in the pilot tunnel were ventilated by the compressed air. For the ventilation of the other working places small centrifugal ventilators of type Capell were used, which could supply 2 m³ of air per second with a length of pipe of 1200 m... There were two each of these on either side, which can be coupled together under pressure. This provisional ventilation was in operation till the final plant was completed." (Schlussbericht, 1914). This meant that this ventilation plant

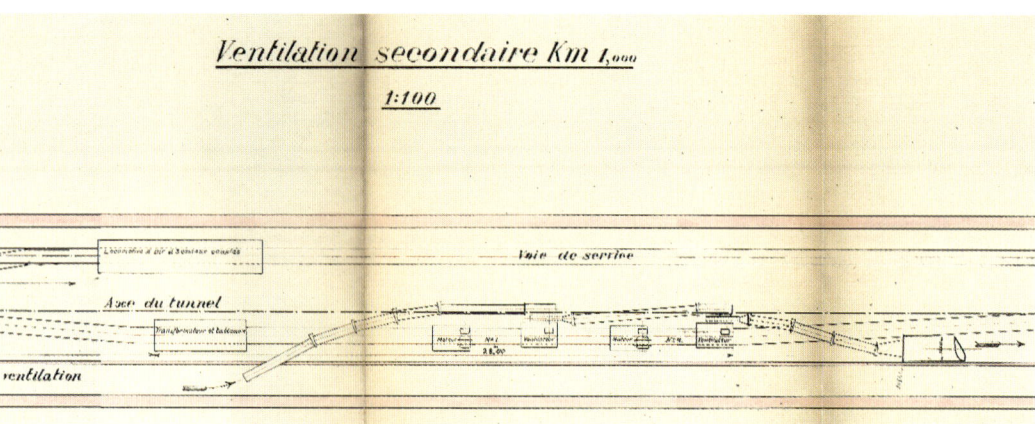

△ For the construction ventilation the fresh air was fed into the tunnel from a separate duct of 6.3 m² cross-section behind a separation wall (Primary ventilation).

▷ Extraction of the air of the primary ventilation system through the fresh air duct for the secondary ventilation system

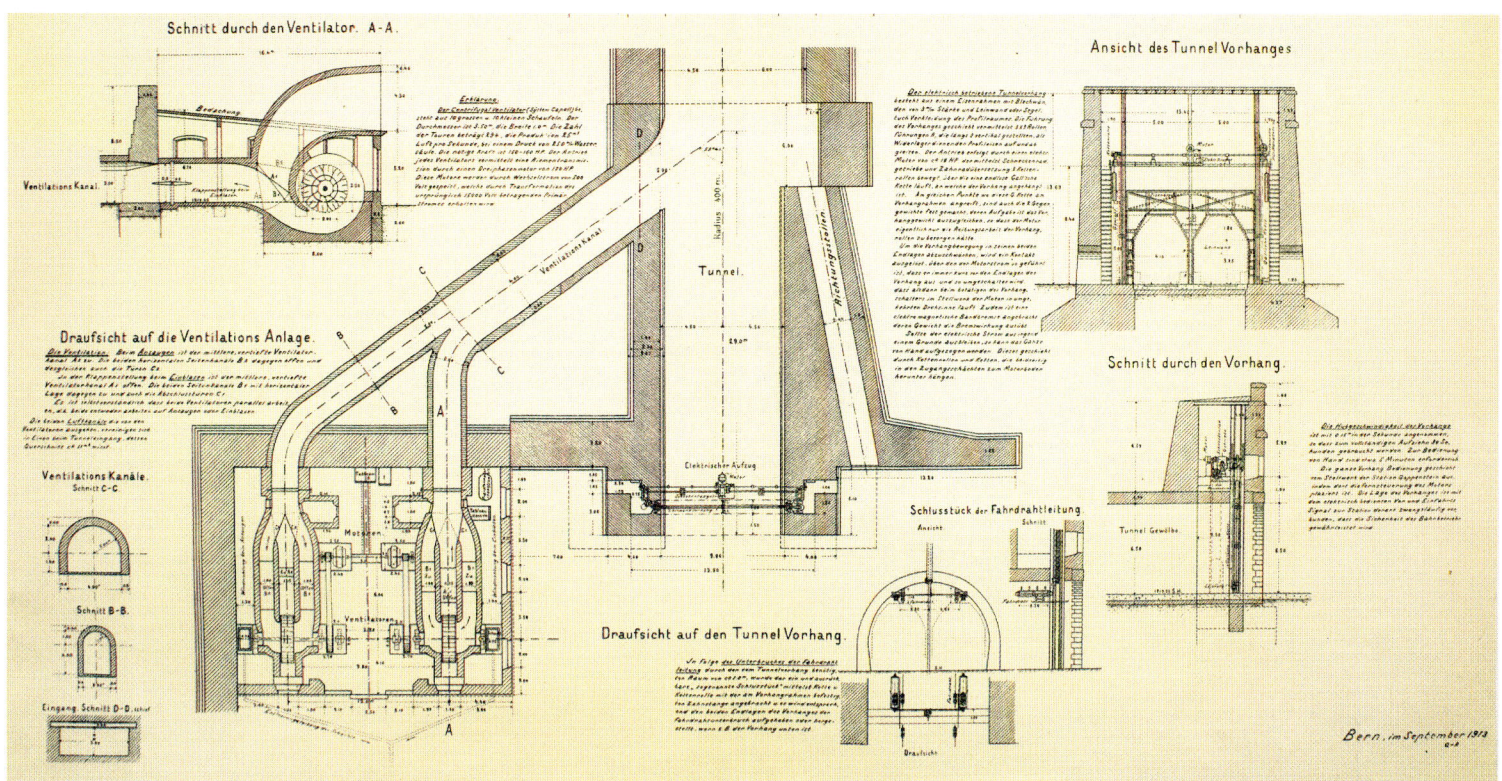

△ Construction ventilation

could be used for the tunnelling work for more than three years.

"On the south side… sprinklers were installed for cooling purposes and to reduce the smoke. … It should be remarked that due to much use of mechanical drilling with drilling machines and rock drilling hammers the air at the working face was tolerable, which was not always the case for the work on the lining."
(Schlussbericht, 1914)

Rock temperatures

"In the Lötschberg tunnel the temperature was measured every 50 m in the drill holes for mechanical drilling. … The maximum rock temperature reached 34.0 °C, which for a temperature increase of 32.4 °C under a maximum overburden of 1490 m represents a geothermal rise of 1 °C for every 46 m. … There was a drop in original rock temperatures up to the end of construction."
(Schlussbericht, 1914)

Operating ventilation

"For the final ventilation on both sides 2 Capell centrifugal ventilators were installed, which supplied 25 m³ of air per second at a pressure of 250 mm head of water. The diameter of the ventilators was 3.50 m. They could only work together for quantity and could not be connected together under pressure. The required performance amounted to 250 h.p."
(Schlussbericht, 1914)

Tunnel Breakthrough and Completion of the Tunnel

Celebration on the occasion of the breakthrough

A big celebration took place on the 14th of May 1911 after the breakthrough had been achieved. Trains filled with guests from Kandersteg (picture bottom left) and Goppenstein (picture bottom right) pass through the whole tunnel, pulled by steam and compressed air engines.

Completion of the tunnels

"The excavation of the full profile was completed on the 31st of March 1912, i.e. exactly a year after the breakthrough of the bottom heading. The masonry lining could be com-

◁ The picture shows the last crew, which achieved the breakthrough.

"On the 31st march 1911 at 3.55 a.m. the breakthrough of the Lötschberg Tunnel took place, whereby both bottom headings 7353 m from the north portal and 7182 m from the south portal met each other." (Andreae, 1940)

932 Der Durchschlag des großen Tunnels.

Hundertein Kanonenschüsse verkündeten am frühen Morgen des 31. März 1911, daß der Durchschlag des Lötschbergtunnels erfolgt sei. Um 8 Uhr morgens brach der Bohrer der Nordseite durch, und eine Stunde später war die Öffnung schon so erweitert, daß Menschen durchkriechen konnten.

Die denkwürdige Episode des Durchschlags schilderte Herr Oberingenieur Rothpletz in folgenden Worten: „Wir hatten immer am Tunneleingang angeschlagen, wieviel Meter noch zu durchbrechen seien. Das war eigentlich unklug. Unter den Arbeitern wuchs die Aufregung mit der Verringerung der Meterzahl. Die Eifersucht, beim Durchbruch dabei zu sein, steigerte sich, und die Leute fingen an zu berechnen, welche Arbeitsschicht es treffen würde. Sie suchten durch das Glück zu beeinflussen, daß sie weniger oder teilweise gar nicht arbeiteten, damit dann der Durchbruch auf ihre Schicht falle. Schließlich mußten wir Leute anstellen, die dafür sorgten, daß überhaupt gearbeitet wurde. Aber nun wuchs auch bei den Ingenieuren die Eifersucht, und es kam dazu, daß wir genaue Verhaltungsmaßregeln für die Ingenieure aufstellen mußten. Und es kam die Stunde, wo der Anschlag besagte, daß nur noch dreizehn Meter zu durchschlagen seien. Nun war die Aufregung allgemein. Wenn sich Hr. Bäschlin (Professor der Geodäsie am Polytechnikum; kontrollierte die Tunnelachse) verrechnet hätte! Offen gestanden, trauten wir seinen Angaben nie recht, und er selbst sagte ja oft, so genau könne man das eigentlich nicht berechnen. Heute bitten wir ihm alles Unrecht reuevoll ab. Donnerstag nachts zehn Uhr hatten wir abgeschossen, voll Erwartung, voll nervöser Spannung. Es war nichts. Da setzte ich mich beiseite, und was mir da durch den Kopf schoß, war sehr ernst. Die Tunnelachse stimmt nicht. Wir werden wieder anfangen müssen, Sondierlöcher vorzutreiben. Um vier Uhr trieben wir ein vier Meter langes Bohrloch vor. Nichts! Schließlich haben wir abgeschossen. Nichts! Hoffnungslos grübelte ich vor mir hin. Plötzlich kommt ein Mann gesprungen. „Durch!" schreit er, „durch!" In dem Augenblick hätte ich am liebsten weinen mögen. Aber mich rief die Arbeit. Dann kam der Augenblick, in dem mir Oberingenieur Moreau an einem Bohrer ein Blumensträußchen von der Südseite her durch das Loch entgegenstreckte. Das sind die schönsten Blumen, die ich meiner Lebtag gesehen habe. Bald darauf kroch Moreau, der beleibte Mann, mit einer erstaunlichen Schnelligkeit durch die Öffnung, und was nun folgte, war ein wildes Durcheinander, ein Trubel und ein Jubel. Arbeiter und Ingenieure, alles trank Champagner, für den meine verehrte Frau gesorgt hatte. Das Ganze ein Bild, das mir unvergeßlich bleiben wird."

△ 22nd April 1912: Ceremony on placing the last keystone in the roof

△ Newspaper cutting on the report of the breakthrough

pleted by the 22nd of April 1912."
(Andreae, 1940)

"The laying of the first layer of ballast for the tracks… was begun on the 5th September 1911 from the north side. … On the 20th July 1912 the laying of the first track was started from the north portal. … After equipping both tracks for electrical operation the first electric engine was driven through the Lötschberg Tunnel on the 3rd June 1913, and on the 6th June the preliminary licensing by the Federal Railways Department took place."
(Andreae, 1940)

Opening of the Lötschberg Tunnel

"On the 28th June 1913 the completion of the Lötschberg Tunnel was celebrated by the Canton Berne with many people taking part."
(Schweiz. Bauzeitung, 1913)

△ The engineers of the construction company on the north side involved in the construction of the Lötschberg Tunnel under the leadership of the chief engineer Ferdinand Rothpletz (Aarau)

The Gotthard Road Tunnel

Total length	16'918 m
Constr. period	10 years
Opening date	5th September 1980
Total costs	686 million SFr.

△ Drilling jumbos at the entrance to the tunnel

History

Overview

1954 Constitution of the Federal Planning Commission

1960 Motion in the Swiss Parliament and in its Upper Chamber. Begin of the planning work of the "Study Group Gotthard Tunnel" under the leadership of the Federal Department of Highways and River Engineering. Proposal for a road tunnel from Göschenen to Airolo. 21st June: Decision of the Federal Assembly to establish the Swiss National Highway Network.

1963 Report of the "Study Group Gotthard Tunnel" concerning the safety in winter of the highway passing through the Gotthard Tunnel.

1965 25th June: Agreement of the Swiss Federal Council to the proposal of extending the National Highway Network to include the Gotthard road tunnel. Formation of the "Commission Gotthard Tunnel" as competent representative of the Federal Department of Highways and River Engineering as well as the Cantons Uri and Ticino.

1968 15th May: Approval of two project variants by the Federal Government. Variant 1 with 4 shafts; variant 2 with 2 shafts and 2 lateral access tunnels. Invitation for tenders: The contractors were obliged to submit a full bid for both variants as well as a contractor's variant with an intermediate attack by means of the shaft at Hospental (divided up into Sections North and South). They were also free to make a proposal for a variant for both projects without an intermediate attack as well as a variant with full face boring for the project with 4 shafts. Further proposals were likewise permitted.
31st October: Submission of tenders.

1969 22nd May: Report and application of the "Commission Gotthard Tunnel" to the Federal Department of the Interior as well as to the governments of the Cantons of Uri and Ticino regarding the choice of the project to be carried out and the award of the contract.
16th June: Choice of the project and approval of the contract awards by the Federal Government. Decision to build an additional safety gallery.

1970 5th May: Official start of construction of the Gotthard road tunnel. Both contractor groups of the Sections North and South began preparations already in autumn 1969.

1972 11th April: Completion of the excavation of the ventilation shaft Hospental.

1973 26th September: Completion of the excavation of the ventilation shaft Guspisbach.

1975 21st February: Decision of the "Commission Gotthard Tunnel" to facilitate an intermediate attack by the shaft at Hospental southwards for the excavation of the safety gallery and the main tunnel.
4th September: Completion of the excavation of the ventilation shaft Motto di Dentro.

1976 26th March: Breakthrough of the safety gallery.
26th July: Completion of the excavation of the shaft Bäzberg.
16th September: Completion of the excavation of the top heading of the main tunnel.

1977 13th May: Completion of the excavation work.

1978 April: Completion of the structural work (concrete lining, intermediate floor and partition wall in the upper part of the cross-section for fresh and used air).

1979 December: Completion of the electromechanical installation work.

1980 Spring/summer: Test of the installations and training of the operating personnel.
5th September: Opened to traffic.

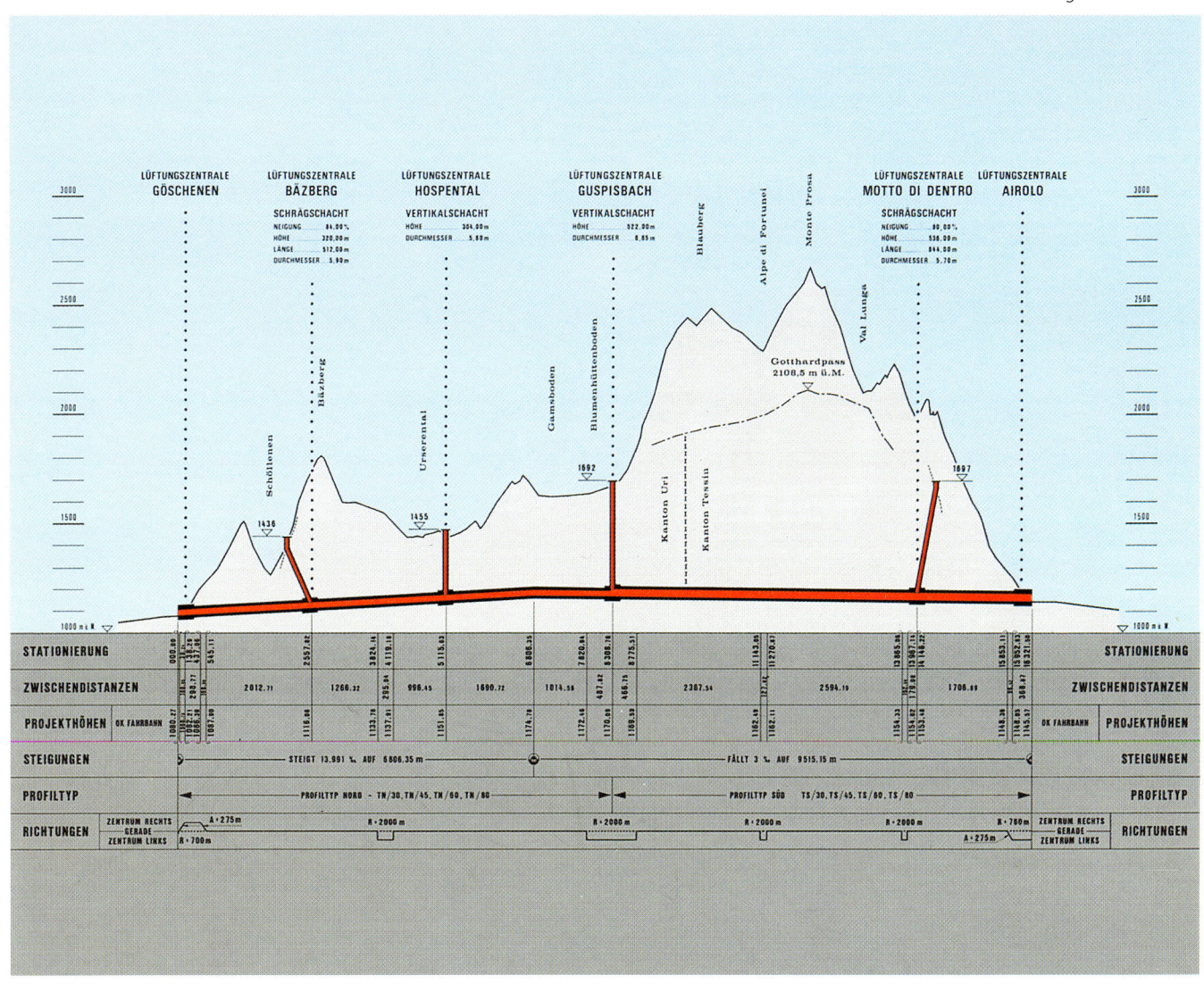

Technical Data and Alignment

Consulting Engineer and construction work

General Supervision
Federal Department of Highways and River Engineering, Berne

Client
Cantons Uri and Ticino

Chief Site Supervision
Section North:
Cantonal Public Works Department Uri
Section South:
Cantonal Highways Department Ticino

Detailed Project
Dr. G. Lombardi, Locarno
Electrowatt Engineering Ltd., Zurich
Dr. A. Haerter, Zurich (Ventilation)

Site Supervision
Section North:
Electrowatt Engineering Ltd., Göschenen
Dr. G. Lombardi, Locarno
Section South:
Cantonal Highways Department Ticino, Airolo

Geological Experts
Dr. T. Schneider, Uerikon (Section North)
Prof. Dr. E. Dal Vesco, Zurich (Section South)

Surveying
Engineering Partnership, Schneider and Weissmann, Chur/Zurich

Contractors
Section North:
Conrad Zschokke Ltd., Zurich
Heinrich Hatt-Haller Ltd., Zurich
Schafir & Mugglin Ltd., Zurich
Ed. Züblin & Co., Zurich
Subalpina (G. Torno & Co.), Lugano
Bau Ltd., Erstfeld
Valentin Studer Ltd., Gurtnellen

Section South:
Walo Bertschinger Ltd., Zurich
Kopp Building Contractor Ltd., Lucerne
Walter I. Heller Ltd., Berne
Rothpletz, Lienhard + Co. Ltd., Aarau
H.R. Schmalz Ltd., Berne

Technical data of the main tunnel

Total length: 16'918 m, of which 16'322 m excavated by mining techniques

Cross-section: Section North: 69–86 m^2
Section South: 83–96 m^2

Gradient: Section North: 1.4 and 0.6%
Section South: 0.3%

Costs

Total costs for planning, building supervision, purchase of land, construction and electromechanical equipment: 688 million Swiss Francs

Of which, extra costs compared to the cost estimate,
Basis 1969: 380 million Swiss Francs (inflation 49%, due to unexpected geological conditions 26%, extensions to project 10%, extra planning costs 3%, added installations 1%, subsequently decision to make immediate attack 5%, pre-investment in a second tunnel tube 6%)

92% of the costs were carried by the Federal Government, the rest by the Cantons Ticino and Uri.

▷ Alignment of the Gotthard Road Tunnel

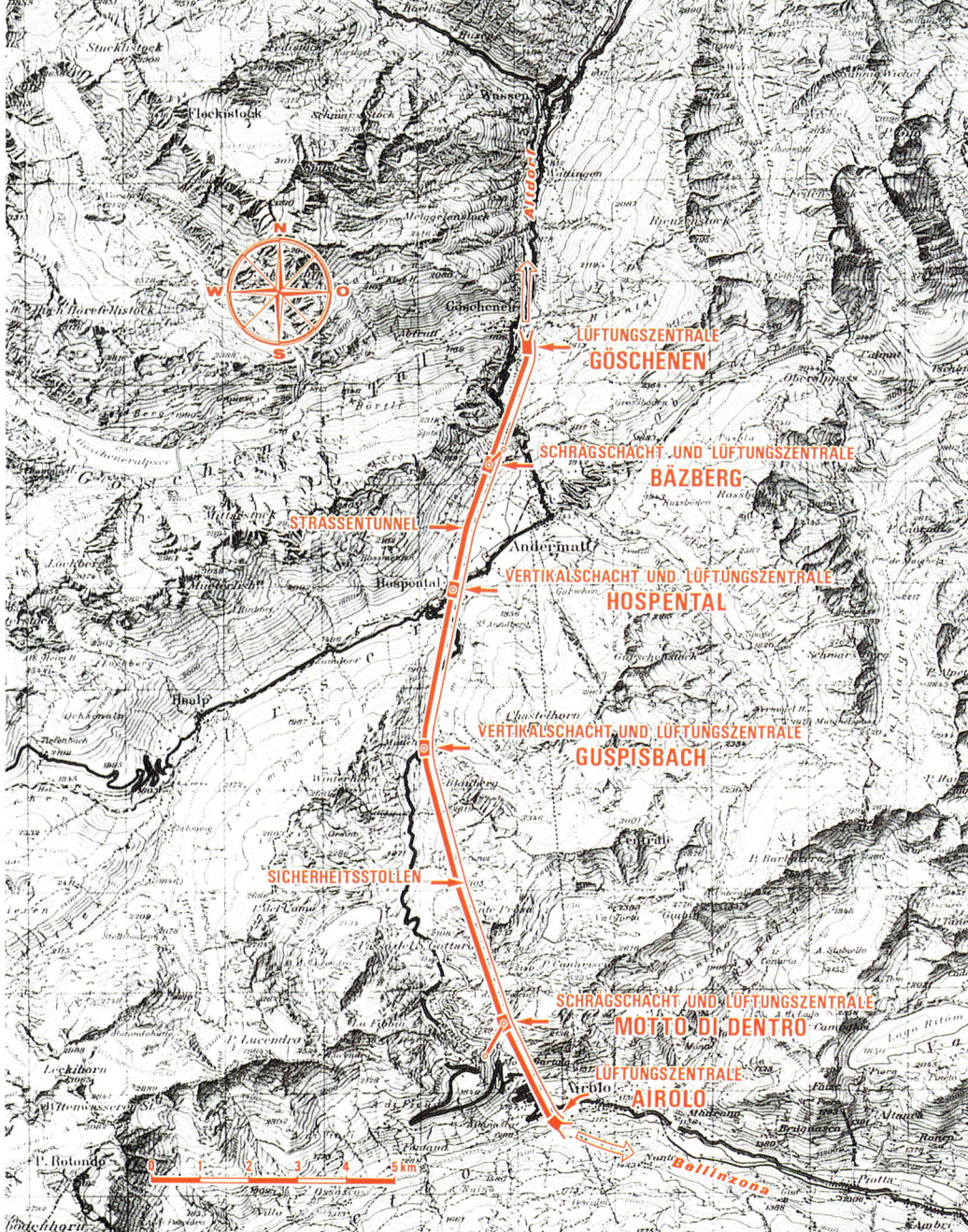

LÜFTUNGSZENTRALE
GÖSCHENEN

SCHRÄGSCHACHT UND LÜFTUNGSZENTRALE
BÄZBERG

STRASSENTUNNEL

VERTIKALSCHACHT UND LÜFTUNGSZENTRALE
HOSPENTAL

VERTIKALSCHACHT UND LÜFTUNGSZENTRALE
GUSPISBACH

SICHERHEITSSTOLLEN

SCHRÄGSCHACHT UND LÜFTUNGSZENTRALE
MOTTO DI DENTRO

LÜFTUNGSZENTRALE
AIROLO

0 1 2 3 4 5 km

Based on economic considerations the tunnel axis was aligned approximately with the cut caused by the river bed of the Gotthard Reuss; thus it deviates from the straight line to the west. The extra length of tunnel was accepted in favour of shorter ventilation shafts driven from the Gotthard pass road. In addition, it was determined from borings that under the valley of Andermatt the bedrock forms a big basin, which extended practically to the level of the railway tunnel. By deviating to the west the bottom of the Urseren gully could definitely be driven under.

Geology

The tunnel passes between Göschenen and Airolo, the southern part of
the Aare massif, the Urseren zone and
the igneous series of the Gotthard
massif. At the north portal first of all
a 160 m stretch of spoil material from
the railway tunnel had to be driven
through using the multiple heading
method. There followed a stretch in
Aare granite; then the tunnel passes
through the southernmost part of the
gneiss zone, which passes with a
sharp transition to the sedimentary
series of the Urseren zone. Here older
rocks were driven through: limestone,
marble, clayey shale from the Jurassic
period, rauhwacke, dolomitic limestone and shale of the Triassic period
and finally shale and sandstone of the
Permo-carboniferous period.

Then comes the Gotthard massif with
sericitic shales in the area of the ventilation shafts at Hospental, para-
gneisses, Gamsboden granitic gneisses and finally in the area of the
south portal the Sorescia gneisses, the
strong water-bearing formations of
the Tremola shales and the geotechnically very unfavourable Triassic formations of the portal area, in which
the tunnel had to be constructed by
the cut-and-cover method.

▷ Horizontal profile, Section North

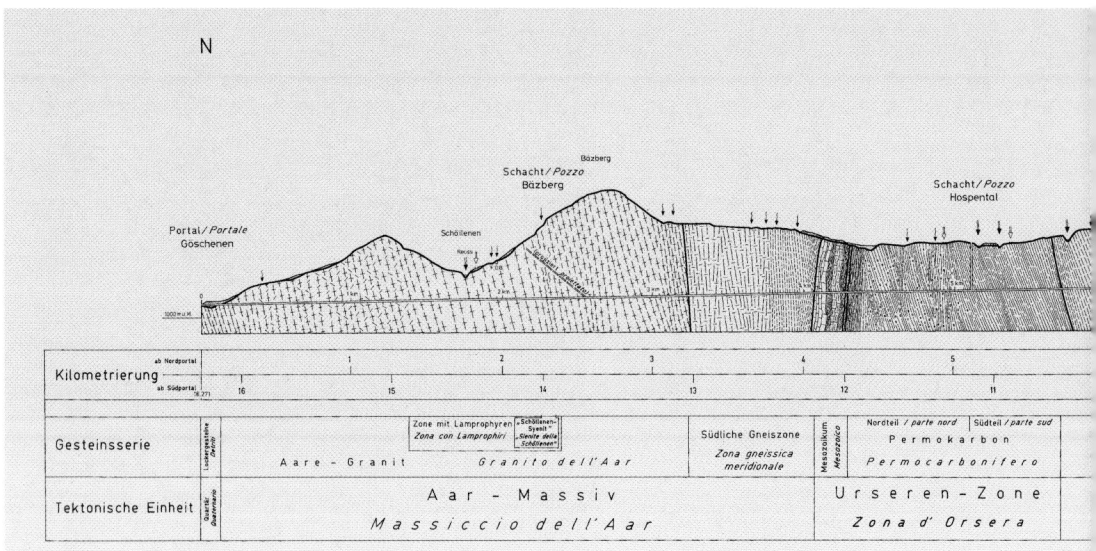

△ Geological longitudinal profile

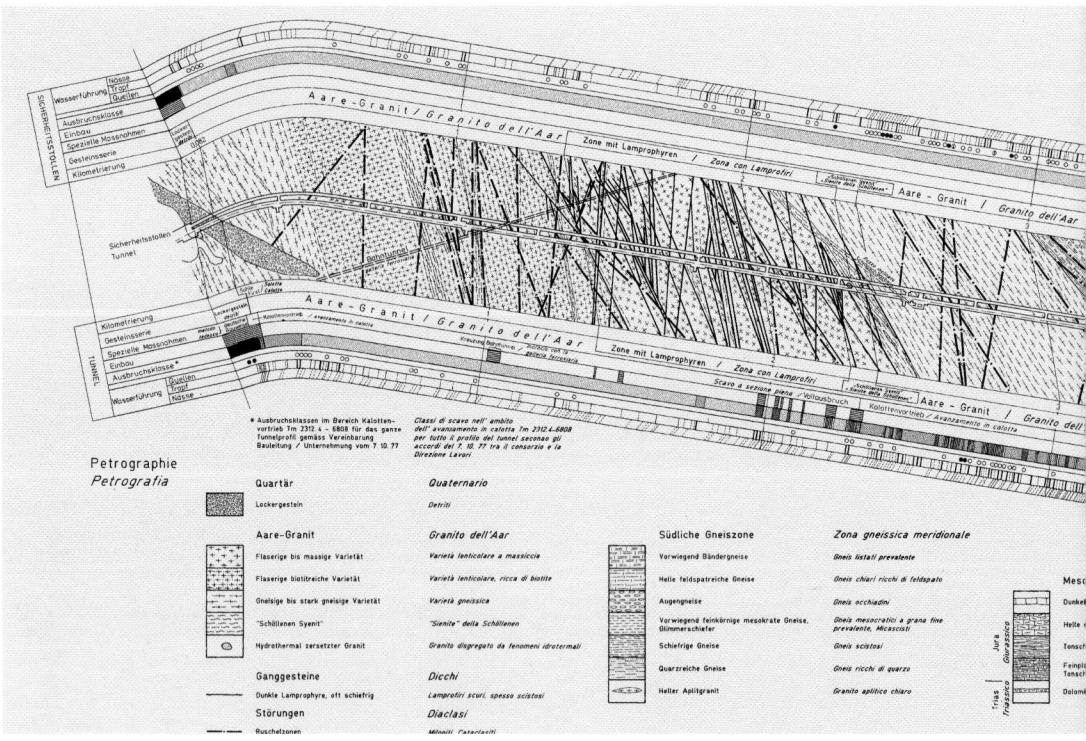

S
URI
TICINO
Monte Prosa
Riale di Fortünei
Val Lunga
Schacht / Pozzo
Motto di Dentro
Riale di Sorescia
Schacht / Pozzo
Guspisbach
ab Nordportal
dal portale Nord
ab Südportal
dal portale Sud
Portal / Portale
Airolo
Voreinschnitt
Trincea di approc
dal Nord
dal Sud
Chilo
Gamsboden - Granitgneis
Gneis Granitico del Gamsboden
Guspis-Zone
Zona di Guspis
Fibbia-Granitgneis
Gneis Granitico della Fibbia
Sorescia-Gneis
Gneis di Sorescia
Rotondo-Granit
Granito del Rotondo
Prato-Serie
Serie del Prato
Pontino-Zone
Zona del Pontino
Sasso-Rosso-Zone
Zona del Sasso-Rosso
Nelva-Zone
Zona di Nelva
Tremola-Serie / Serie della Tremola
Trias Triassico
Moräne Morena
Nufenen-Zone
Zona della Novena
Quartär Quaternario
Gotthard — Massiv
Massiccio del San Gottardo

Durchschlag Sicherheitsstollen
foratura del cunicolo di sicurezza
Lotto Nord
Los Nord
Los Sud (Attacco intermedio Hospental)
Los Süd (Zwischenangriff Hospental)
Lotto Sud
Los Süd
Cunicolo di sicurezza
umidità
stillicidio
sorgente
Infiltrazioni d'acqua
Classe di scavo
Armatura
Provvedimenti speciali
Unità geologica
Chilometraggio dal Nord
gneissica meridionale
Mesozoikum
Mesozoico
Permokarbon / Permocarbonifero
Paragneise des Gotthard-Massivs / Paragneis del Massico del S. Gottardo
Gamsboden-Granitgneis / Gneis granitico del Gamsboden
Cunicolo di sicurezza
Galleria principale
Schacht / pozzo
Guspisbach
Schacht / Pozzo
Hospental
Galleria stradale
Classi di scavo
Zona gneissica meridionale
Kalottendurchschlag
perforazione attraverso calotta
Mesozoico
Calcari scuri
Calcari chiari marmorei, marmi
Argilloscisti con intercalazioni fini di calcari
Calcari tabulari p.p. con intercalazioni di argilloscisti
Dolomia e gesso
Permokarbon
Weissliche zersetzte Serizitschiefer bis Phyllite
Helle lagige Serizitschiefer
Konglomeratische chlorit- und biotitführende Serizitschiefer
Biotitführende Chlorit-Serizit-Schiefer und -Gneise
Idem mit Bänken aus hellen quarz- und feldspatreichen Gneisen (z. T. vulkanogen)
Permocarbonifero
Scisti alternati chiari da sericitici a filladici
Scisti a sericite chiari laminari
Scisti conglomeratici a sericite con clorite e biotite
Scisti e gneis a sericite e clorite con biotite
Idem con banchi di gneis chiari ricchi di quarzo e feldspato (p. p. vulcanogenici)
Paragneise
Feinkörnige mesokrate Zweiglimmergneise
Idem mit Serpentinlinsen
Idem mit Linsen und Bändern aus Amphibolit
Paragneis
Gneis mesocratici a due miche
Idem con lenti di serpentine
Idem con lenti e letti anfibolitici
Gamsboden-Granitgneis
Heller porphyrischer granitischer Gneis
Gneis granitico del Gamsboden
Gneis granitico chiaro, porfiroide

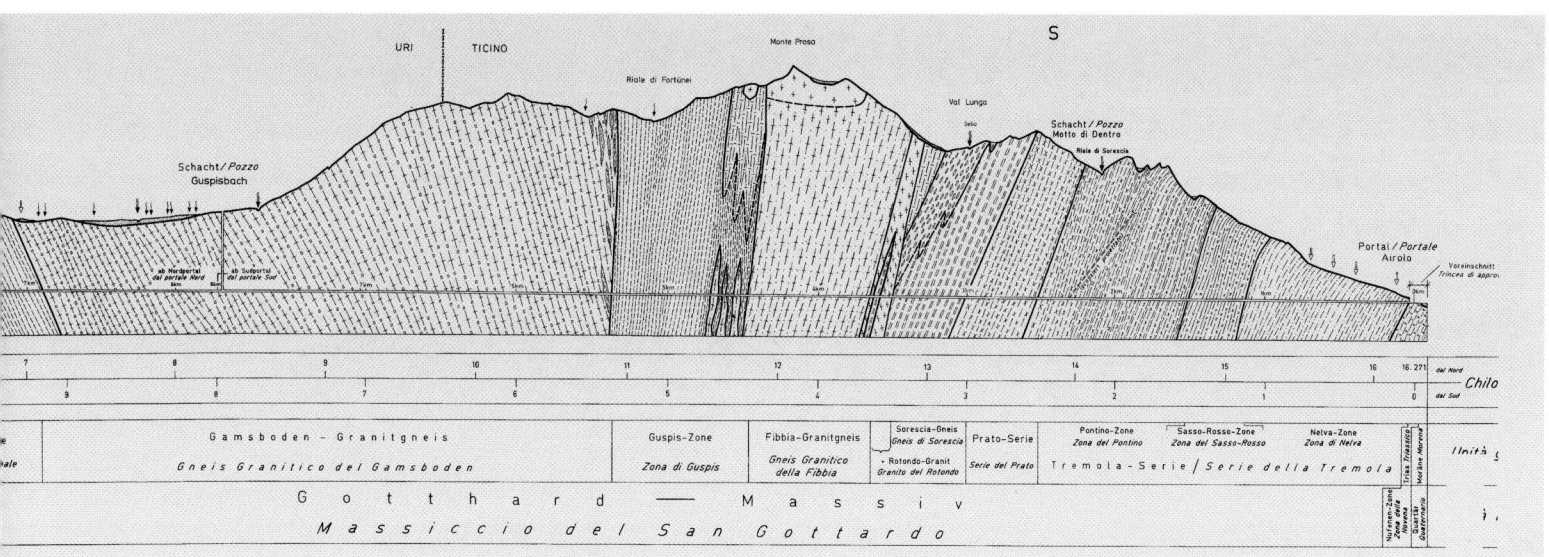

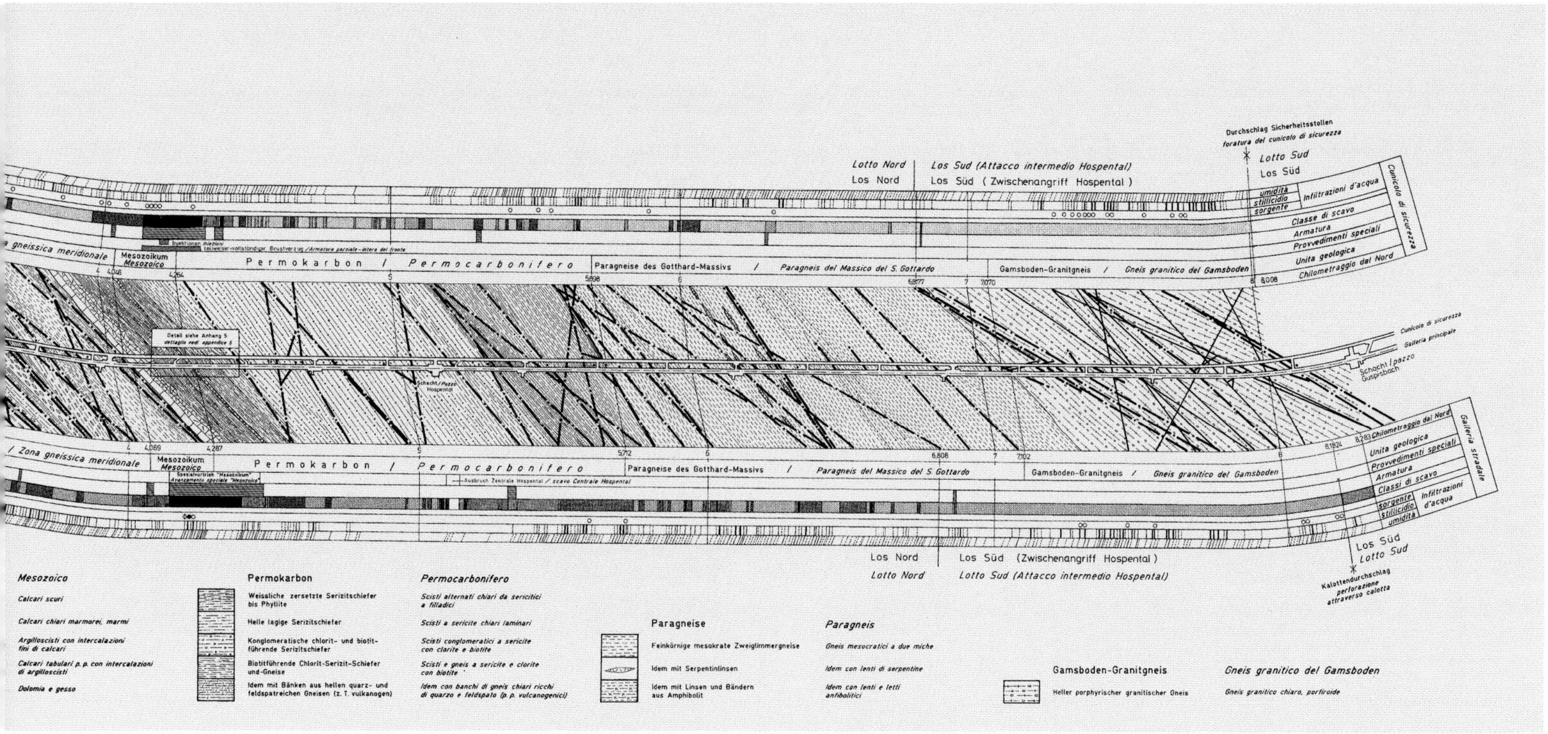

Construction Installations and Programme

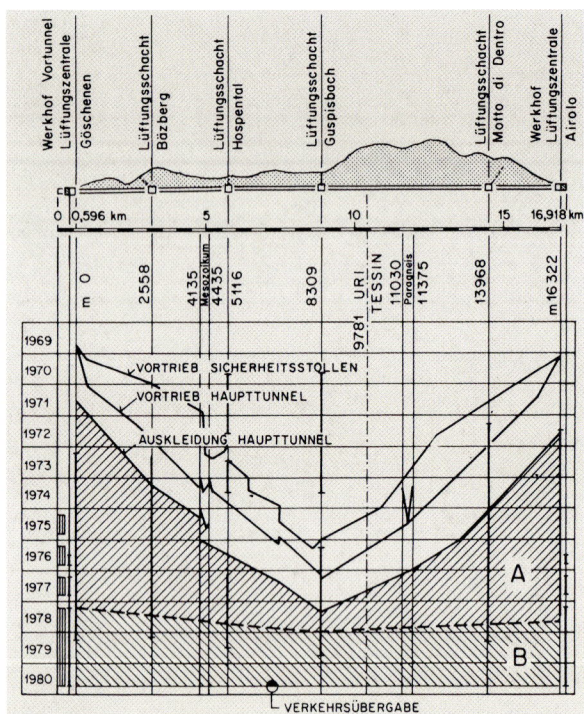

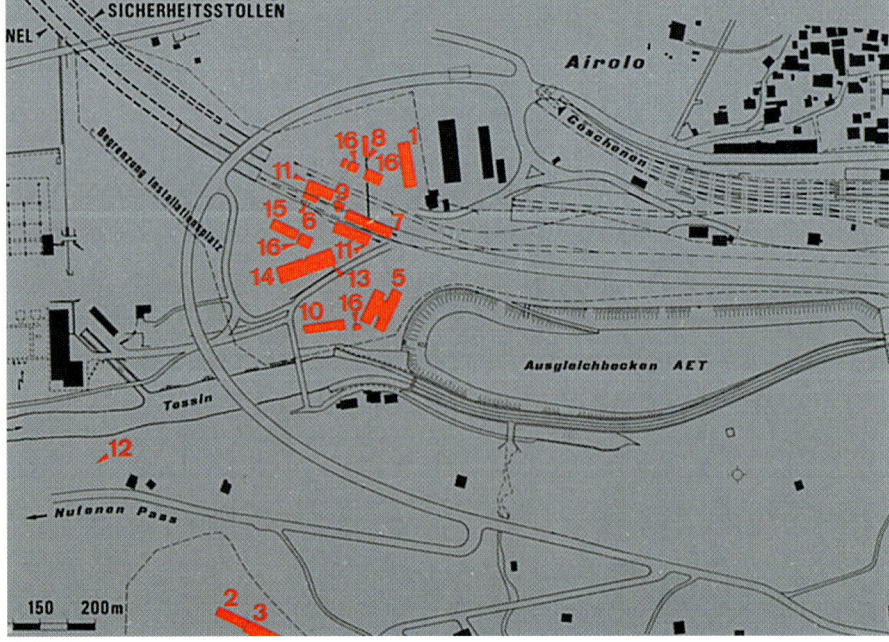

△ Construction programme and excavation sequence

Facilites for the construction installations in Airolo:
- 1 Site office
- 2/4 Accommodation
- 3 Wash rooms
- 5 Canteen
- 6/7/8 Silos
- 9 Concreting tower
- 10 Metal working shop
- 12 Explosives
- 13/14 Workshop
- 15 Compressors
- 17 Materials handling plant

Facilities for the construction installations in Göschenen:

1/10	Site office
2/4	Accommodation
3	Sanitary area
5	Canteen
6/7/9	Silos
8	Site office
11	Concreting tower
12	Metal working shop
13	Explosives storage
17	Workshop
18	Compressors
22	Materials handling plant

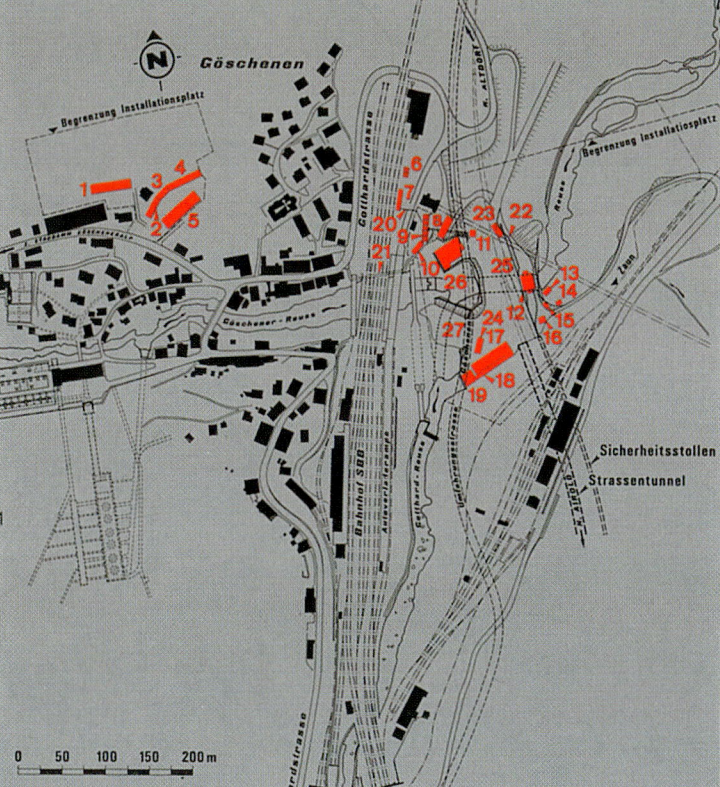

Normal Cross-Sections of the Tunnel, Driving of Safety Gallery

The normal cross-section for the tunnel traffic is 7.80 m wide and a clear 4.5 m high. The cables for power supply, telecommunications, control and hydrants are located beneath the sidewalks. In the space between the tunnel wall and the tunnel lining are the secondary fresh air ducts and other non-operating cables. In the space between the intermediate roof and the tunnel roof are the ducts for fresh and exhaust air.

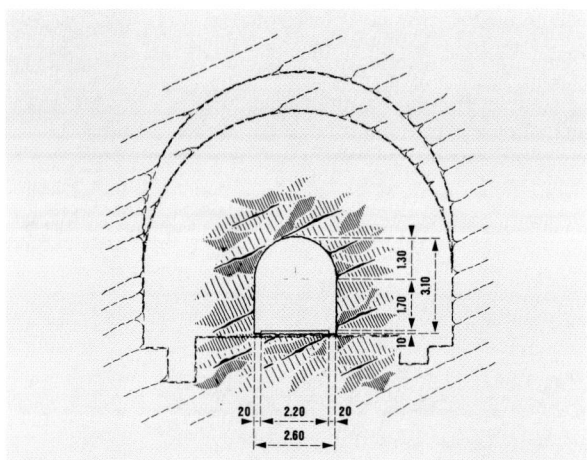

◁ Normal cross-section for the safety gallery in the Section North. The distance between the axes of the safety and the main tunnels is 30 m.

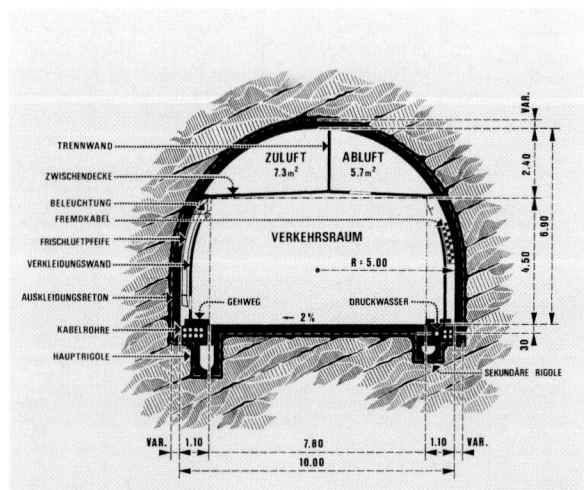

△ Section North

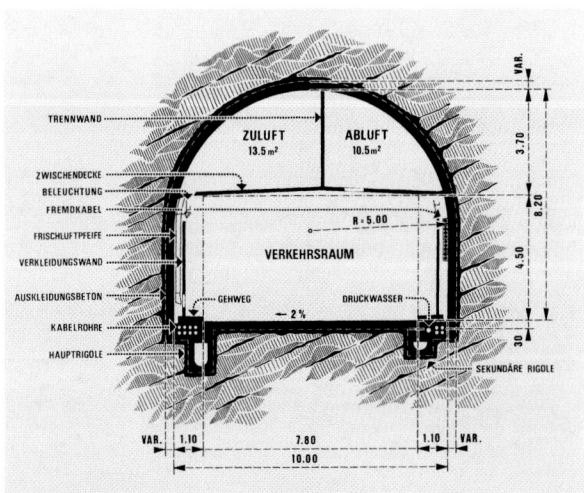

△ Section South

△ Difficulties in excavating the safety gallery in the paragneiss squeezing rock zone under high water pressure in the Section South

△ Deformed or badly yielding steel supports due to the rock pressure in the safety tunnel in the region of the paragneiss

▷ In the foreground the reconstruction may be seen (horse shoe or circular profiles), in the background the badly yielding steel rib supports in the Section South

The safety gallery was driven at a certain time interval before the main tunnel. The work in this area indicated that there would be difficulties when constructing the main tunnel. At the working face in the Section South sturdy round timber sets were needed to support brittle rock. The groundwater could escape by means of drainage holes.

Excavation of Section North

△ Control of profile in main tunnel using laser beams

△ Partial excavation in the passing place below the rail tunnel, Section North

△ Left drilling jumbo platform at the working face, full face excavation

△ Drilling jumbo platform Ingersoll-Rand for the full face excavation

▷ Sketch of the full face excavation method in the Section North

Excavation of Section North Mesozoic Formation

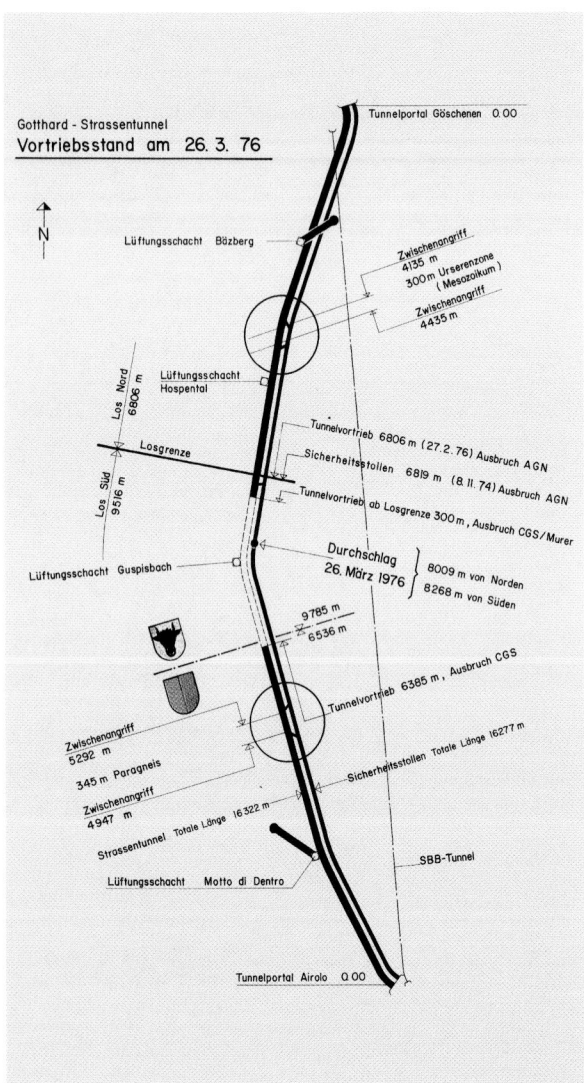

◁ Intermediate attack in the paragneiss, Section South, and in the Urseren zone (Mesozoic rock, Section North). Progress to 26th March 1976. Breakthrough of the safety tunnel

In the geotechnically poor formation of the Mesozoic rocks, Section North, the tunnel was driven by the multiple drift method. Excavation sequence: side headings, excavation of the roof heading by means of forepoling, excavation of the centre and concreting of the invert. Average rate of excavation: 1.45 m/day

▷ Cross-section for partial excavation

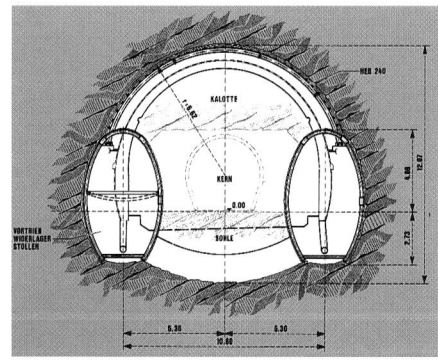

▽ Excavation working sequence

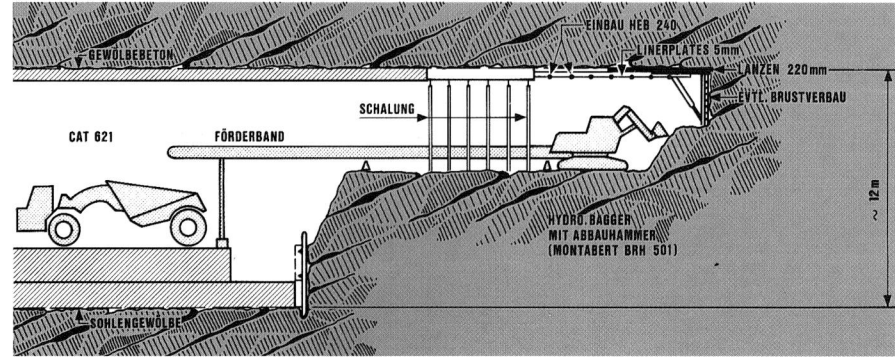

▷ Excavation by means of forepoling in the heading

▷ Excavation of the centre core with an excavator. On the sides the two side headings. Loading the debris into the rubber tyre vehicle

Concreting the Section North

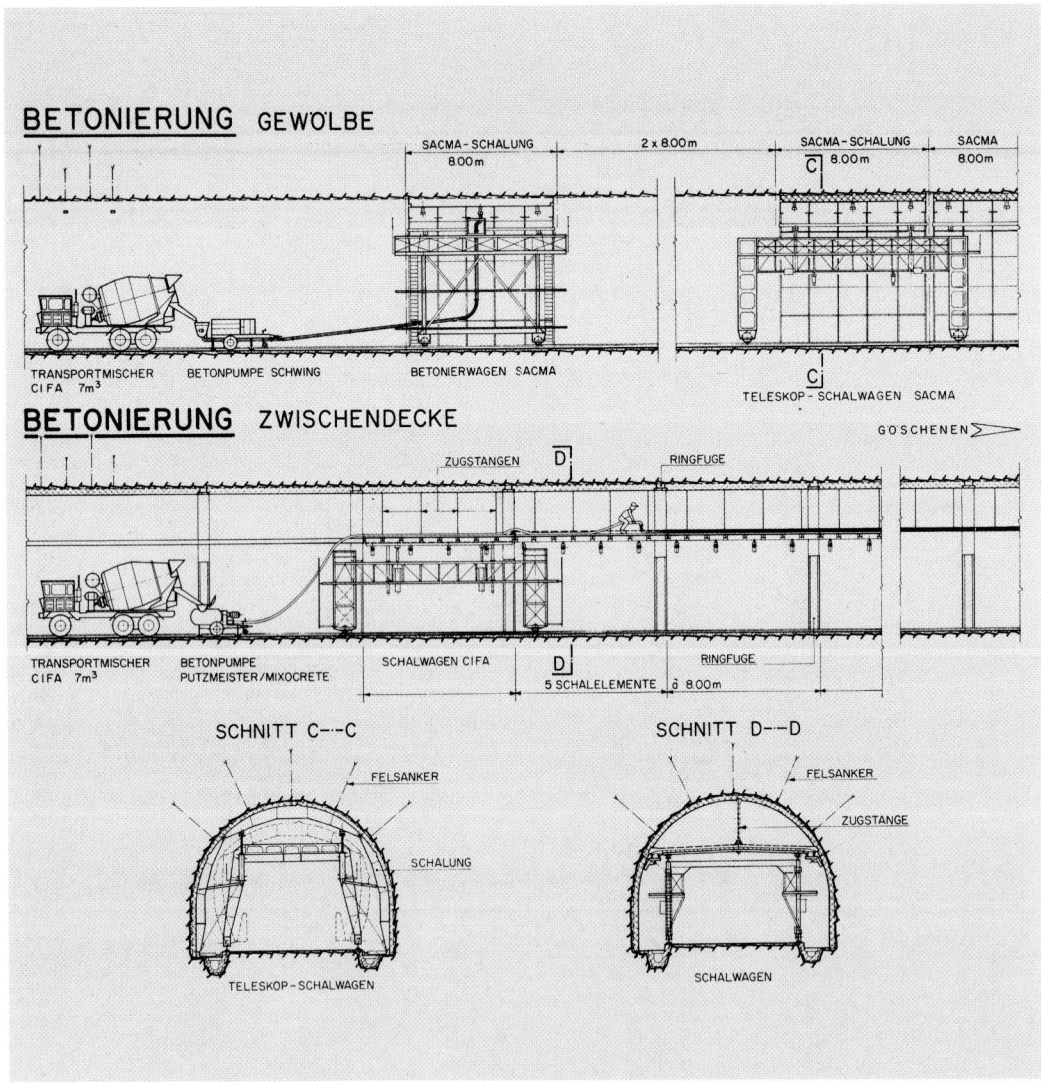

BETONIERUNG GEWÖLBE

SACMA-SCHALUNG 8.00 m   2 x 8.00 m   SACMA-SCHALUNG 8.00 m   SACMA 8.00 m

C

TRANSPORTMISCHER CIFA 7m³   BETONPUMPE SCHWING   BETONIERWAGEN SACMA

C

TELESKOP-SCHALWAGEN SACMA

BETONIERUNG ZWISCHENDECKE

GÖSCHENEN ➤

ZUGSTANGEN   D   RINGFUGE

TRANSPORTMISCHER CIFA 7m³   BETONPUMPE PUTZMEISTER/MIXOCRETE   SCHALWAGEN CIFA   D   RINGFUGE

5 SCHALELEMENTE à 8.00 m

SCHNITT C--C

FELSANKER

SCHALUNG

TELESKOP-SCHALWAGEN

SCHNITT D--D

FELSANKER

ZUGSTANGE

SCHALWAGEN

◁ Sketch of the concreting operation for the tunnel roof and the intermediate floor, Section North, main tunnel

▷ Assembly of the tunnel forms for placing the concrete lining at the material handling area North. Concreting plant, tunnel forms and tunnel forms transport wagon

△ Sealing and tunnel lining (forms)

△ Concreting the intermediate floor in the main tunnel

Section South
Sliding Floor Method

For the first time in Europe the "sliding floor method" was used in the Section South. On a 240 m long, 5–9 m wide sliding steel floor consisting of 5 elements, the marshalling area for changing the railway wagons was pushed forwards section-wise using hydraulic presses. Therefore, to load the transport wagons near the working face no reinstallation was necessary.

◁ The removal of the spoil material was carried out in the Section North using rubber tyre vehicles, in the Section South using rail wagons. Cat 980 during mucking

▽ The sliding floor method. Longitudinal section

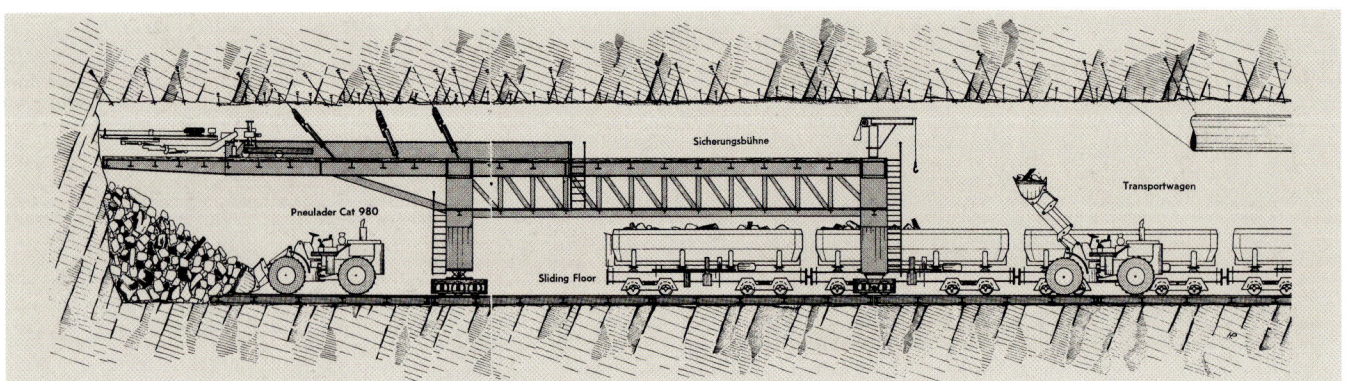

▷ Sliding floor before being covered over

▽ Operation sequence for the sliding floor

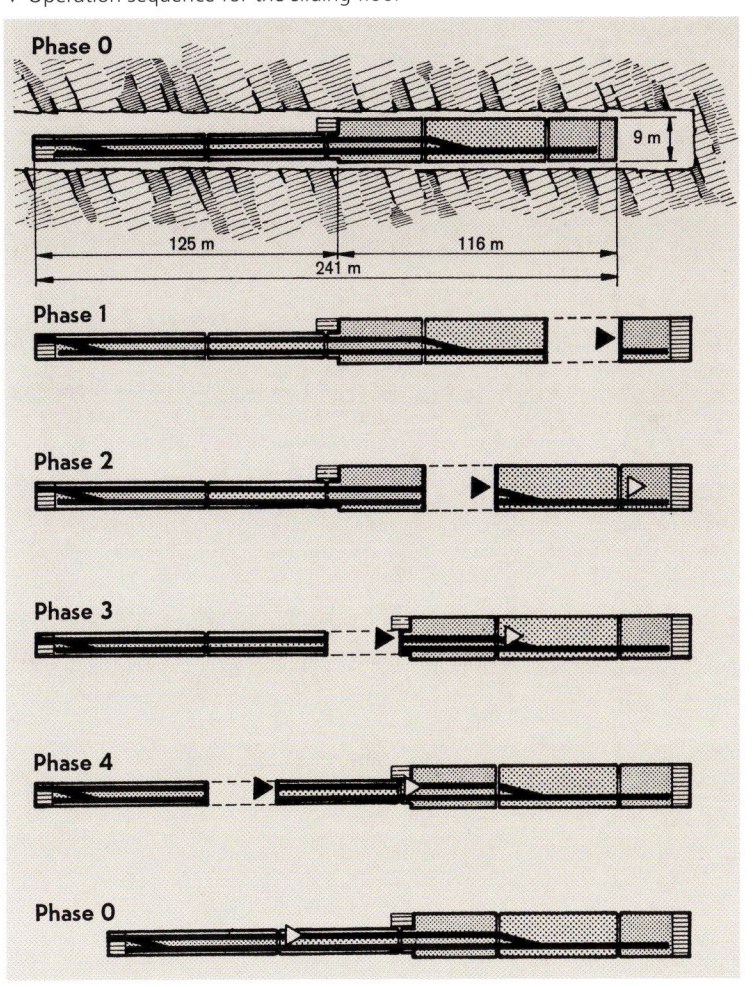

Phase 0

125 m   116 m

241 m

9 m

Phase 1

Phase 2

Phase 3

Phase 4

Phase 0

Ventilation

22 fresh air ventilators are installed in 6 ventilation rooms. Automatic control equipment controls the necessary amount of air. The fresh air is fed into the tunnel from the portals and shafts through the inlet duct and blown sideways into the traffic space. The used air is sucked up to the discharge duct above the intermediate floor every 16 m.

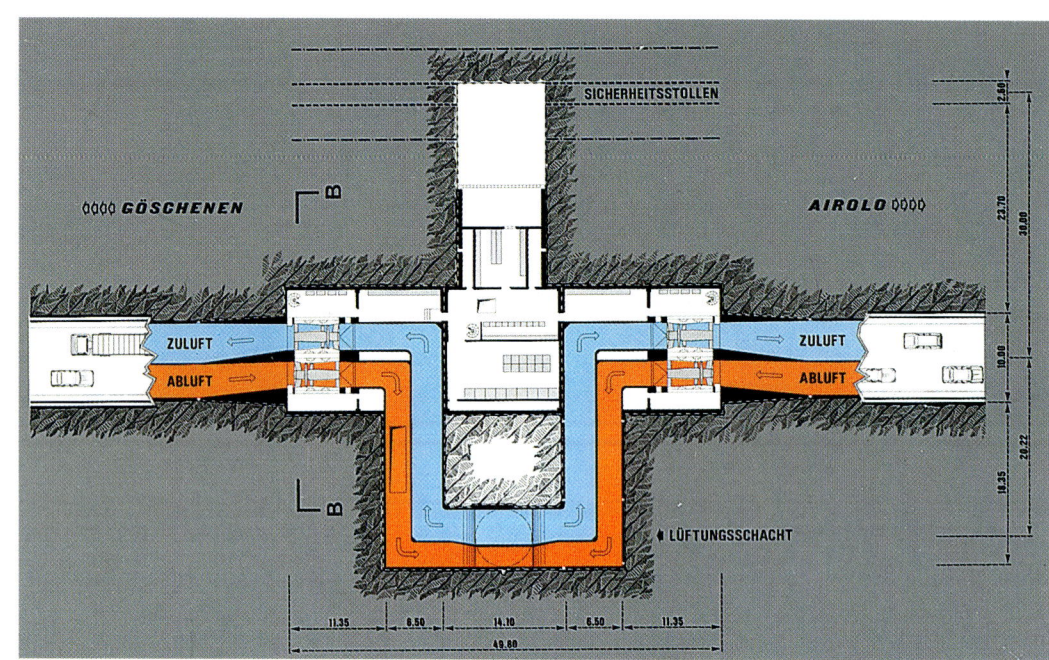

△ Sketch in plan of the ventilation rooms

▽ Sketch longitudinally of the ventilation system

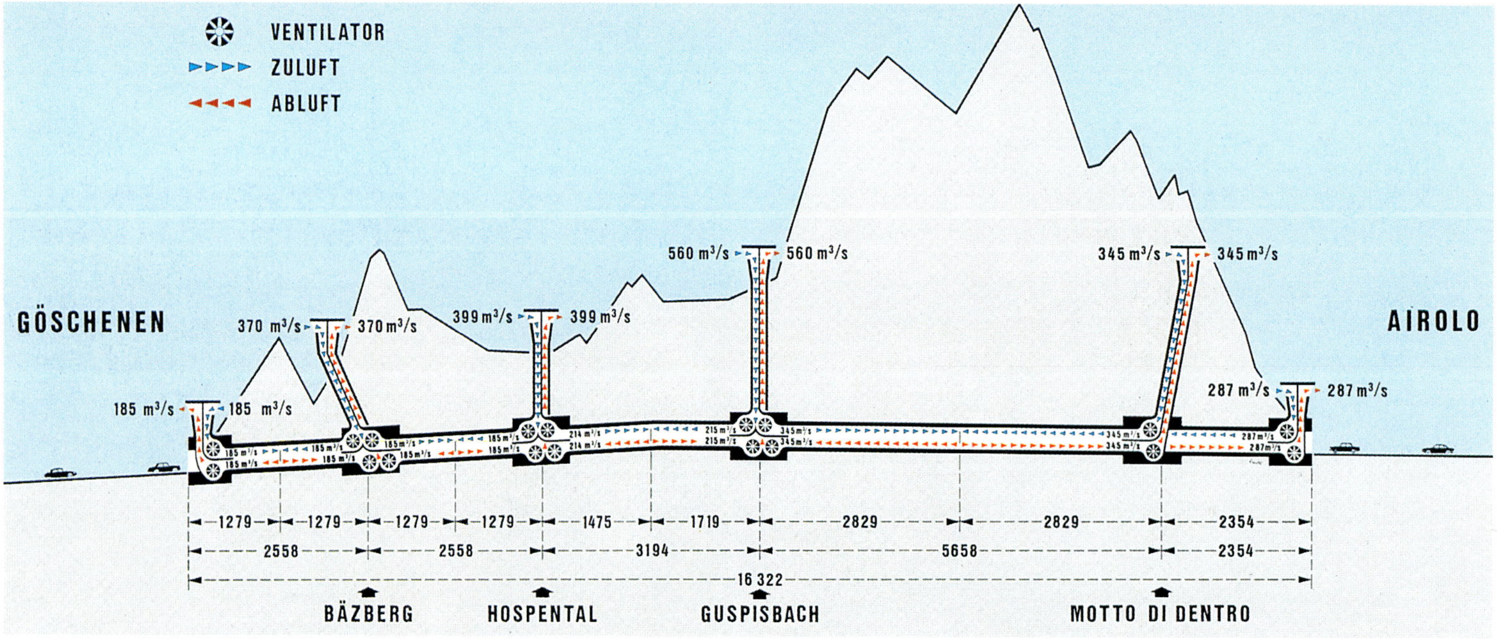

▷ Inlet and discharge ducts above the traffic space. The intermediate floor has not yet

▽ Ventilators

▽ Inlet and discharge ventilators in a ventilation room

Ventilation Structures Shafts

The two inclined ventilation shafts Motto di Dentro and Bäzberg were excavated using a Tunnel Boring Machine (TBM) of 3 m diameter from bottom to top. Afterwards the shafts were enlarged to their final diameter using a larger machine working in the opposite direction.

The two vertical shafts Hospental and Guspisbach were sunk in their final diameter in one operation from top to bottom. After sealing with a plastic foil, the shaft walls were concreted as well as the partition walls between the intake and discharge air using the slip form construction method from bottom to top.

▷ Plastic sealing in the inclined shaft Motto di Dentro. The shaft walls and the partition wall have not yet been built.

▷ Inclined shaft Motto di Dentro. The head of the boring machine breaks through the ground surface.

▷ Enlarging the diameter of the sloping shaft Motto di Dentro

△ Mechanical cutting machine (diameter: 3 m) in the sloping shaft Motto di Dentro

▷ The vertical shafts Guspisbach and Hospental were drilled with a triple boom drilling machine and blasted. In the picture is the drilling tower and excavation tower Hospental.

A Comparison of the Alpine Tunnels in Switzerland

	Gotthard Rail Tunnel	Simplon Tunnel I + Parallel Tunnel	Lötschberg-Tunnel	Gotthard Road Tunnel
Length (m)	14'982	19'803	14'605	16'322
Average section (m²)	56.8	49.6+9.0	57.6	101.2
Volume of spoil (m³)	847'670	990'000	836'858	1'651'000
Volume of masonry and concrete (m³)	156'000 masonry	280'000 masonry	186'542 masonry	370'000 concrete
Total construction period (months)	111	96	66	124
Rate of excavation (m/day)	4.47	6.9	7.33	5.7 m
Total construction costs (Swiss Francs)	66'666'581.–	78'000'000.–	50'289'300.–	686'000'000.–
Construction costs/tunnel metre (Swiss Francs)	3910.–	3940.–	3459.–	42'030.–
Total amount of explosives used (kg)	1'400'000	1'520'000	960'918	2'819'000
Explosives used per m³ of rock (kg/m³)	1.65	1.54	1.14	1.71
Error in breakthrough alignment: in the direction (cm)	30	20	25.7	5
in the height (cm)	5	9	10.2	6
in the length (cm)	710	100–200	41	5
Maximum No. of personnel	3874	3420	3250	ca. 700
Total No. of working hours		70'000'000	27'457'680	9'262'000
Working hours per tunnel metre(h) and per m³ of excavation (h)		3500 70	1890 33	567 5.6
Fatal accidents	177	67	64	17

Documents and photographs

Financial Support

Bauamt des Kantons Uri, Altdorf,
H. Bargähr

Bern Alpenbahn-Gesellschaft, BLS,
Bern

B. Fantoni, Bauunternehmung, Brig

Gotthard-Strassentunnel,
Werkhof Airolo

Institut für Geodäsie, ETH Zürich

Kreisdirektion II der SBB, Luzern,
Bahnarchiv, E. Luginbühl

Rothpletz, Lienhard+Cie AG, Aarau,
P. Rothpletz

W. Scheidegger, Ambri

Dr. T. R. Schneider, Geologe, Uerikon

Schweizerisches PTT-Museum, Bern

Stadtmuseum Aarau, Dr. Weingarten

Veramess Engineering, Aarau,
Dr. H. Aeschlimann

Rothpletz, Lienhard+Cie AG, Olten

Financial support was given by the
following organisations:

Engineering firms, different
branches of the building industry as
well as public authorities, the Swiss
Federal Government and the Federal
Institute of Technology Zurich (ETH)

Literature

Andreae, Ch. (1916): Einige Erfahrungen im Lehnenbau an der Südrampe der Lötschbergbahn, Schweiz. Bauzeitung, 6. Mai

Andreae, Ch. (1940): Baugeschichte der Lötschbergbahn (Claude Jeanmaire, Lötschbergbahn im Bau, 75 Jahre BLS, Archiv Nr. 58, Verlag Eisenbahn, 1989)

Baeschlin, F. (1911): Über die Absteckung des Lötschbergtunnels, Schweiz. Bauzeitung, Band 58

Besondere Bestimmungen der SBB (1902) für die Ausführung des Mauerwerks für Unterbauarbeiten, Bern

Briger Anzeiger (1899), Mitteilung, Juli

Berner Alpenbahn (1907): Die Dienstbahn von Frutigen nach Kandersteg, Sonder-Abdruck aus der Schweiz. Bauzeitung, Band L, Nr. 21, 23. Nov.

Eclog. Geol. Helv. (1904), VIII, November

Moeschlin, F. (1957): Wir durchbohren den Gotthard, Artemis, Zürich

Pestalozzi, S. (1904): Die Bauarbeiten am Simplontunnel, Sonder-Abdruck aus der Schweiz. Bauzeitung, Zürich

Pressel, K. (1906): Die Bauarbeiten am Simplontunnel, Schweiz. Bauzeitung, Band XLVII

Rapport (1909) de la commission des experts désignés par l'entreprise générale du chemin de fer des Alpes bernoises, 8 janvier

Rosenmund, M. (1905): Über die Anlagen des Simplontunnels und dessen Absteckung, Jahresbericht der Geographisch - Ethnographischen Gesellschaft Zürich

Rothpletz, F. (1944): Erinnerungen an die schwersten Tage meines Lebens und deren Folgen, 23. März

Schlussbericht (1914) über den Bau des Lötschbergtunnels der Berner Alpenbahn

Schweiz. Bauzeitung (1913), Mitteilung, 5. Juli

Steiner-Ferrarini, M. (1992): Wahlheimat am Simplon, Verlag zur alten Post, Brig

Vom Bau des Simplontunnels 1898–1921 (1921), Zur Feier der Schlusssteinlegung im Tunnel II, 4. Dezember

Wyss-Niederer, A. (1979): Sankt Gotthard – Via Helvetica

Zollinger, A. (1910): Berner Alpenbahn, Bern–Lötschberg–Simplon, Schweiz. Bauzeitung, 25. Juni

Zum Durchschlag des Lötschbergtunnels (1911), Schweiz. Bauzeitung, 8. April